Computers, Education and Special Needs

E. PAUL GOLDENBERG
SUSAN JO RUSSELL
and
CYNTHIA J. CARTER
with
SHARI STOKES
MARY JANE SYLVESTER
and
PETER KELMAN

ADDISON-WESLEY
PUBLISHING COMPANY

Reading, Massachusetts
Menlo Park, California
London · Amsterdam
Don Mills, Ontario · Sydney

Intentional Educations, Series Developer
Peter Kelman, Series Editor
Richard Hannus, Cover Designer

This book is in the
Addison-Wesley Series on Computers in Education

Opening photographs for Chapters 1, 2, 3, 5, 6, and Resources by E. Paul Goldenberg.
Chapter 4 opening photo by Marshall Henrichs.

Figs. 1.1, 1.2, 1.4, 1.5, 1.6, 5.4, and 5.5 by Marshall Henrichs.
Fig. 1.3 by Martha Lester Goldenberg.
Figs. 1.4 and 3.1 by Grafacon.
Fig. 2.1 by Design Technology.
Figs. 2.2 and 2.3 by E. Paul Goldenberg.
Fig. 2.4 by Visualtek.
Fig. 5.3 by Harvard Associates.

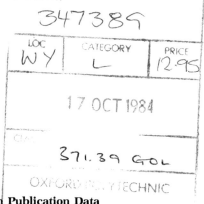

Library of Congress Cataloging in Publication Data
Goldenberg, Ernest Paul.
   Computers, education, and special needs.

   (Addison-Wesley series on computers in education)
   Includes index.
   1. Computer-assisted instruction.   2. Handicapped
children—Education—Curricula.   I. Russell,
Susan Jo.   II. Carter, Cynthia J.   III. Title.
LB1028.5.G59   1984   371.9'043   83-21487
ISBN 0-201-05778-6

Copyright © 1984 by Addison-Wesley Publishing Company, Inc.

All rights reserved. No part of this publication may be reproduced, stored in a retrieval system, or transmitted in any form or by any means, electronic, mechanical, photocopying, recording, or otherwise, without the prior written permission of the publisher. Printed in the United States of America. Published simultaneously in Canada.

ISBN 0-201-05778-6
ABCDEFGHIJ-AL-8987654

Computers, Education and Special Needs

Foreword

The computer is a rich and complex tool that is increasingly within the financial means of schools. Like any educational tool, it comes with inherent advantages and disadvantages, is more appropriate for some uses than others, is more suited to some teaching styles than others, and is neither the answer to all our educational ills nor the end of all that is great and good in our educational system. Like any tool, it can be used well or poorly, be overemphasized or ignored, and it depends on the human qualities of the wielder for its effectiveness.

The purpose of the Addison-Wesley Series on *Computers in Education* is to persuade you, as educators, that the future of computers in education is in your hands. Your interest and involvement in educational computer applications will determine whether computers will be the textbook, the TV, or the chalkboard of education for the next generation.

For years, textbooks have dominated school curricula with little input from classroom teachers or local communities. Recently, television has become the most influential and ubiquitous educator in society yet has not been widely or particularly successfully used by teachers in school. On the other hand, for over one hundred years the chalkboard has been the most individualized, interactive, and creatively used technology in schools.

Already, textbook-like computerized curricula are being churned out with little teacher or local community input. Already, computers are available for home use at prices comparable to a good color television set and with programs at the educational level of the soaps. If teachers are to gain control over computers in education and make them be their chalkboards, the time to act is now.

Each book in the *Computers in Education* series is intended to provide teachers, school administrators, and parents with information and ideas that will help them begin to meet the educational challenge computers present. Taken as a whole, the series has been designed to help the reader:

- Appreciate the potential and the limits of computers in education.
- Develop a functional understanding of the computer.
- Overcome apprehension about and fear of the computer.
- Participate in efforts to introduce and integrate computers into a school.
- Use the computer creatively and effectively in teaching and administration.
- Apply a discerning and critical attitude toward the myriad computer-related products offered in increasing volume to the education market.
- Consider seriously the ethical, political, and philosophical ramifications of computer use in education.

Practical Guide to Computers in Education is the basic primer for the series. *Computers, Education and Special Needs* is one of a number of books in the second tier of the series, each dealing with computer applications in particular educational contexts. Others include *Computers and Reading Instruction*, *Computers in Teaching Mathematics*, and *School Administrator's Guide to Computers in Education*. Still other titles are planned for this part of the series, including books on computers and writing, business education, science, social studies, and the elementary school classroom. Each book in this second tier picks up where the *Practical Guide to Computers in Education* leaves off. Each is more focused and provides far more practical detail to educators seriously considering computer use in their schools and curricula.

Computers, Education and Special Needs is perhaps the most important volume in the series. It is my hope that not only educators, but parents, legislators, and organizations for the handicapped will read it, be inspired by it, and take appropriate action. Computers present unprecedented opportunities for "normalizing" the lives of many whose differences from the mainstream have resulted in unnecessarily limited educational experiences. This book describes some of the ways in which computers can broaden the horizons of students with special needs, but it is only the beginning. It is up to people like those who are reading this book to take up the baton and run with it. Good luck. We hope we have provided you with an inspirational, yet practical, start.

Peter Kelman

Preface

We had actually done a considerable amount of the planning—even some of the writing—of this book before we articulated to ourselves what critical educational issues were appropriate to a book on computers, education, and special needs. For a while, we had divided up the educational experience, and the techniques we would talk about, along traditional subject lines—mathematics, language arts, music and art, etc. But then, as we noticed that we were not about to write the "etc.," we began asking ourselves why not. It was clear how language arts, for example, would be a special concern in a book like this; but how would teaching of social studies or science or mathematics to handicapped students differ from one another? The techniques of communication and manipulation of materials that would be needed to make the material accessible are no more specific to social studies and science than to any other subject area. This is, of course, not to say that there are no pedagogical issues. There is a lot to say, for example, about making manipulatives accessible in a science course, but the special issue is the accessibility. Good science teaching, once that problem is in the background, would then be what it always is.

There were also decisions to be made about definitions and the scope of the book. For example, it is clear that among those individual needs that require special consideration, foreign language background is one. Yet, it is both awkward and misleading to classify as disabled a child whose only difference from the common culture is fluency in another language. Still, although the situations of the bilingual student and the cerebral palsied student are rather different, the contribution that the computer can make to each student can be quite similar. We decided to include both, mindful of the sensitivities of those who point out that language knowledge is anything but a handicap. In Chapter 2 we develop other important definitions and delineate the themes dealt with in the remainder of the book.

Communication is a necessity for learning and for life. Regardless of one's intelligence, a lack of fluent, flexible communication cuts one off from the activities that feed learning. The inability to read well—whether it arises from visual impairment, perceptual disorder, lack of fluency with the language, or educational disadvantages—limits access to information. Thus Chapter 3 deals with this issue of language and communication.

Then there is the issue of access to content. There are the physical limitations of some children that interfere with their access to content. Further, the difficulties that we, the special educators, face often leads to the watering down of curricula for special students. Chapter 4 selects mathematics as an example, but is intended to illustrate the richness with which any subject may be approached. Without significant opportunity and reasonable expectations, learners have little chance of significant learning.

Motivation and its affective relatives, including the sense of one's own control and autonomy, form a third key issue, especially for those whose autonomy is constantly being threatened by both reality and prejudice. This is the focus of Chapter 5.

These three issues—communication, access, and motivation—are central to learning and to this book.

Of course, there are also various practical matters including assessment (discussed in Chapter 6) and the actual adaptations possible with the new technology (discussed in Chapter 7).

The book begins in Chapter 1 with a dream that is already within reach technologically. It ends in Chapter 7 with some administrative realities of implementing that dream—giving concrete examples of how such special programs may be set up, from funding to classroom design.

This book provides pointers to what has been done, and is intended to help you select from these resources approaches that will work in your setting. Each of us brought to this book our own experiences and philosophies. Meeting as a group, we designed what we believe to be a coherent whole that is superior to the sum of its parts. Still, each of us takes primary responsibility for the chapters on which we were the principal authors: E. Paul Goldenberg and Cynthia J. Carter for Chapters 1, 3, and 5; Susan

Jo Russell for Chapters 4 and 6. Chapters 2 and 7 were the results of collaboration among all of us, Chapter 2 having been organized by Peter Kelman out of contributions by Shari Stokes, Paul Goldenberg, Cindy Carter, and Peter Kelman, and Chapter 7 having been organized by Paul Goldenberg and Cindy Carter out of writing by themselves, Susan Jo Russell and Mary Jane Sylvester. Credit for conceiving the project, assembling this group of authors, leading the initial brainstorming of ideas, and keeping the group on task goes entirely to Peter Kelman whose perseverence, sometimes even under protest from the rest of us, was remarkable.

There are too many people to whom we owe intellectual debts, so we will mention only those who helped us directly with this book. However, it should be clear that none of these people bears any responsibility for the views expressed herein. We would like especially to thank Brian Harvey whose good advice, clever insights, and strong criticism (sometimes momentarily devastating) greatly improved this book. Others whose criticisms and suggestions were very helpful to us in the revision process included Dr. Glenn Kleiman, Dr. Sylvia Weir, Judy Wilson, and Veronica Kenney. We also thank Marilyn Martin, Carol Nuccio, Helen Pollack, Laura Koller, and various staff members from Intentional Educations for their contributions, particularly for organizing the resources in this book. In addition, several members of the Addison-Wesley staff—most directly involved were Peter Gordon, and Marshall Henrichs—worked closely with one of the principal authors especially during the final stages of production, giving considerable time, skill, and advice. To them, and to Elydia Siegel of Superscript Associates, special thanks is due.

Finally, there are the children, most of whom we cannot name, who have been our most important source of inspiration for this book.

October, 1983

E. Paul Goldenberg
Susan Jo Russell
Cynthia J. Carter
Shari Stokes
Mary Jane Sylvester
Peter Kelman

Contents

1 Computer Supported Learning Environments 1

2 Education and Special Needs 19

    Learning and Education 21

    Special Needs in Education 21

    Motivation 22

    Access to Information 24

    Communication 28

    The Tasks of a Teacher 30

    Possible Roles for the Computer 32

    Conclusions 42

3 Communication and Language Arts 45

    Choosing Appropriate Computer Interventions—Looking at the Disability 50

    Choosing Appropriate Computer Interventions—Looking at the Student's History 56

    Language-Learning Environments 59

    The Language-as-Currency Model 63

    The Language-as-Subject-Matter Model 71

    Tools for Communication 81

    Conclusions 85

4 Mathematics: A Case Study of Computers in the Curriculum 87

How Much Mathematics Is Enough 89

Drill and Practice 93

The Woods Are Lovely, Dark, and Deep: Beyond Drill and Practice 95

A Walk through the Forest 113

Finally, Guideposts and Warnings 118

Discussion of Problems 119

5 Motivation and Autonomy 123

Internal Motivation 123

Learned Passivity 125

Computers, Music, and Art 128

Computers and Motivation 132

Engaging the Unmotivated 134

Widening Horizons 139

Making Competence More Attractive Than Incompetence 140

Conclusions 146

6 Assessment 149

Assessment and the Curriculum 151

Computer Enhanced Assessment 152

Understanding Children's Thinking 156

Why Do I Have to Take This Stupid Test?: Assessment as Communication 161

Information Overload: Computers as Managers 164

Conclusions 167

7 Getting Started 169

Becoming the School Computer Expert 169

Customizing the Computer for Individual Needs 181

One School's Experience 188
Epilogue 192

Glossary 195

Resources 205

Index 261

Computers, Education and Special Needs

Computer Supported Learning Environments 1

You have just walked into the special learning center at Ellis Central School. In many ways, it seems quite a familiar setting, except, perhaps, for such a large presence of computers. The room is lively and attractive, full of people involved in a variety of activities. Some children are working in groups; others by themselves.

Near the door there are several art projects underway: a boy fingerpainting at a low table, a girl designing a city scene on a TV screen by typing instructions on a computer keyboard, and a girl in a wheelchair getting set up to draw on a computerized **graphics tablet.** A young man is working with two children creating a brightly colored and beautifully textured abstract painting. Their materials are not the traditional tempera and paper, but computer and large TV screen. An older woman is sitting nearby with an artist's pad, making a charcoal sketch of one of the children at work.

You ask about the girl in the wheelchair. The teacher explains that 14-year-old Tammy has severe **spastic-athetoid cerebral palsy** and, until recently, has been thought to be quite retarded. Her reading skills had been barely at first-grade level, and even those were acquired with great frustration and discouragement. But reading is especially important to Tammy, the teacher continues, because she cannot speak at all. Tammy depends on printed representations of language for communication. She must be taught to compose messages by selecting letters, words, and phrases from "menus" that appear on the computer screen. By pointing to various locations on the same tablet that she is preparing to draw on, she will indicate her selections and assemble them into a message. She will learn to use the same motor skills for communication as she is now using for freehand drawing.

You are struck by the absurdity of her current task. How can freehand drawing make sense for a child whose motor control is so terribly poor that she can hardly even point? How could she do more than just scribble? How, in fact, can she even hold the pen?

An aide straps the pen to Tammy's hand, and as you watch her hand flail about on the tablet with no apparent pattern, you see also the seriousness with which Tammy is working. As she moves, a very recognizable "T" begins to emerge on the TV screen in front of her. She pauses, obviously pleased, and begins drawing the second letter of her name.

What about the other children in the art area? Are there unusual stories behind their work? Well, of course, you don't really know a child until you can tell stories that make that child seem absolutely unique, but, no. Mostly, these kids are here because they like art, and because they and their teachers have planned this to be art time. Nobody here has a story quite as dramatic as Tammy's. But the teacher tells stories, anyway.

Fig. 1.1 Unfiltered representation of Tammy's hand movements as they are recorded by the graphics tablet she uses for freehand drawing.

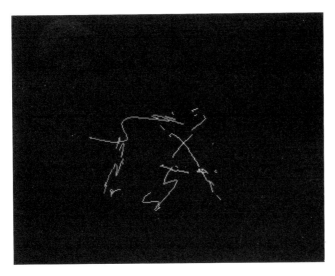

Fig. 1.2 When Tammy's movements are properly processed, it becomes more apparent that she is printing her name. The letters TA which were only hinted at in Fig. 1.1 are readily seen here. With practice, Tammy's writing becomes even clearer.

The young man working with the children is a professional artist, exploring the computer medium for the first time. The older woman sketching is a member of a senior volunteer program and spends three days a week in the learning center. During her breaks, she beautifies the room with her sketches of children at work (Fig. 1.3). The girl working with the city scene has been a timid and poor student in mathematics classes, but is, in the course of her art work, using and developing the same principles of mathematics here that she had found hard to learn in class.

The reading corner appears positively full. A slight child of perhaps 12 is sitting on an aide's lap listening to him read and frequently interjecting comments about the story. Four students are sitting at a table, two working hard in what appear to be workbooks, the other two typing on the keyboard of "The Talker" and listening as the little blue box next to the keyboard speaks out loud. Tammy uses a device somewhat like "The Talker" to speak for her. The children here are using it to experiment with the sounds of letters alone and in combination. Sometimes they type gibberish. Even then, "The Talker" tries to pronounce what they type. More often, they seem to be trying to get "The Talker" to say particular words. As you approach, "The Talker" says, "What is your name?" and all four children

Fig. 1.3

at the table giggle. The girl curled up in a beanbag nearby continues reading without seeming to have noticed.

Another three children in the reading area are huddled over an engrossing computerized fantasy game of adventure and daring. One boy seems to be in charge, having the coveted central position at the keyboard. The monitor screen in front of him shows a brightly colored picture with the caption "Near the tree there is a unicorn with a golden ring." He reads the caption out loud to the others who excitedly urge him to take the ring. The boy slowly types, "Take the ring" and the computer responds with a picture captioned, "The unicorn smiles and offers you a ride." You are told that this boy had been lagging way behind in reading since the beginning of school, but loves this game and has been advancing to more and more sophisticated text each time he plays. His teacher periodically rewrites the game with new twists to the adventure, new treasures and dangers—and new vocabulary. The computer program makes it easy to tailor and personalize the story, including editing the text, plot, and pictures.

You notice yourself thinking about what is not present. The room might have been dominated by students sitting at computer terminals

Fig. 1.4

and being taught programmed lessons at their own pace or playing noisy arcade-like games. So far, neither seems to be the case. The work (or is it play?) that the students are engaged in is of clear educational value, but hardly like a programmed learning sequence. Student involvement is intense, but not reminiscent of the arcade addicts. You make a mental note: it is the teachers and students who are in control here. The machines do not initiate, guide, teach, or manage anything. The teachers are guiding and teaching, the students are taking the initiative, and the computers respond. But so cleverly and in such interesting ways! The teacher has obviously taken others on tours through the center and responds to your expression with a chuckle.

Probably the biggest surprise to come out of this room, you are told, is Tammy's ability to draw—and what that has revealed about her knowledge, learning potential, and motivation. But another child's story is almost as dramatic. That boy was "so retarded that, at 12, he still could not name his colors." His relatively mild motoric impairments were enough to jeopardize his communication, and vastly confuse assessment. Before the computer enlarged the variety of activities in which he could engage, there had been little that he could do to demonstrate his intelligence, and nobody had found a reliable way to show that his ability to match, manipulate, and name shapes, for example, was totally unimpaired. A subsequent test revealed that his failure with colors was simply due to color blindness!

Across the room, back near the art area, you notice a boy and a girl, both perhaps 13 years old, wearing earphones, apparently concentrating intently on what they are hearing, and occasionally typing on a keyboard. You ask the teacher what they are doing. With an expression mixing real pleasure with amusement, the teacher just waves a hand toward the pair, as if to suggest that you should ask them directly. You approach the children and notice a book lying open next to the girl. On the right-hand page is a picture of a musical instrument that looks something like a fur bagpipe; on the left is a short paragraph and three separated selections of very simple-looking music, with lyrics in Hungarian! The other child, a boy, looks up first and you ask him what they are doing. His answer is brief and direct: they are studying music. You ask more. As the boy senses your genuine interest, his style changes. Articulately, and in far greater depth than you could have expected, or probably wanted, he explains some of the details of music theory that he is studying, slipping in mention of his own musical compositions and his years of piano playing. The girl, he adds, is doing a project with Hungarian folk music, trying to learn enough about its characteristics to compose new pieces that "sound Hungarian." "On the computer?," you ask. "Listen," he says, and hands you the earphones. "I composed that. The sound quality is not very good, but it can play better than me. Well, not really better (it has no feeling!), but it doesn't need to practice. And it doesn't have to worry about where its fingers are, either!"

You glance back at Tammy. The connection is so obvious. It doesn't matter that Tammy's poor dexterity is labeled "a handicap" and that no special label is given to deficiencies in the finger skills of these talented young musicians. The computer is providing exactly the same service for all three. The young musician practically said so. The computer is being their fingers, making it easier for them to use their minds.

The computer activities are showy: they are, after all, what attracted you to visit this school. But as you glance back at the areas you have visited, you are impressed by how many people are not working with computers. Styles of learning with and without the computer blend as naturally here as do the different projects themselves. You wander on.

In a pod in the middle of the room are four students sitting in front of keyboards and screens. One of them is making a drawing on the computer by typing commands. Two are doing some kind of computational drill. Ah, so the school does have some traditional computer assisted instructional programs! The fourth is doing subtraction problems working alternately with the computer and in a teacher-made workbook. When he wants help with a workbook problem, he can ask the machine.

Do all the children in the school squeeze through this one tiny mathematics area? Four computers and a table? In reality, it turns out that there is no mathematics area at all! Many students, like the girl drawing the cityscape back in the art area, engage in important mathematical acitivities in the course of other projects. The purpose of this special pod is simply to be small and separate.

Pointing to a corner territory that you haven't yet visited, the teacher explains that it had originally been set aside and largely enclosed by sound baffles for students who didn't want the distraction of the activities in the other areas. It had been used for awhile by students who were just beginning to learn about computers, but as the machines became familiar objects around the school, fewer and fewer students seemed to want the isolation that area had been designed to provide. Eventually, the corner territory became a writing laboratory, and the four computers that originally had been in the corner were set up here in the center of the room, less visually isolated than the writing lab is, but still separate enough from the other areas to provide some privacy. The use of this central pod varies. At the moment, as you had noticed, three of the four students are working on arithmetic activities, so it looked like a mathematics area. The student who is drawing usually spends his time in the art area, and may have come here just because the art area was full.

Your attention is drawn back to the boy doing subtraction. His machine, unlike a calculator, does not simply show the answer, but actually demonstrates a method for solving the problem, by explaining out loud how to work it out on paper. Sometimes it shows only the standard paper-and-pencil method, but sometimes it shows the method graphically. Right now, the machine is showing the boy how to subtract 30 from 624. It sets up a fantasy in which it helps the boy manage his own ping-pong ball factory. There are 624 ping-pong balls in stock, and an order comes in for 30 ping-pong balls to be shipped out. The machine represents the

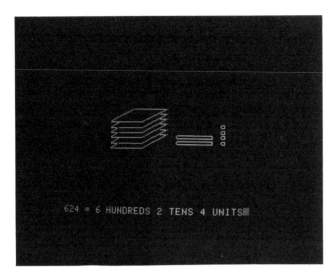

Fig. 1.5a Sometimes the program represents the ping-pong balls graphically in their packages...

Fig. 1.5b ... and sometimes the representation is more symbolic.

624 ping-pong balls as 4 individually wrapped ping-pong balls, 2 strips of 10 ping-pong balls, and 6 packages of 10 strips.

It explains that the order for 30 ping-pong balls will be shipped out as 3 strips. Since there are only 2 strips on the shelves, the boy must unwrap one of the 6 packages to get more strips. As the computer shows the package being unwrapped and its contents (10 strips) being stacked with the other 2 strips that were already shown, it also shows the "borrowing" operation in standard arithmetic notation off to the right. Now, the 3 strips can be shipped out, leaving 5 packages, 9 strips, and 4 ping-pong balls—594 ping-pong balls—in stock. The machine talks through each step, and repeats steps as the boy asks for them.

You walk over to the one remaining area, protected by its sound baffles. A small cluster of students is talking with a teacher. The teacher waves you in to join them and comments that this is the writing lab. The group is discussing ideas for writing short poems. There are several computers in the area, but nobody is using them. After listening for a while, you ask if they ever do use the computers. A smallish boy pipes up over the

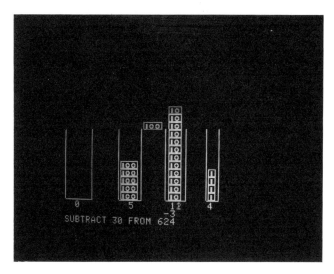

Fig. 1.6 One package of 100 ping-pong balls is removed from the hundreds bin and hovers between the hundreds and tens bins while strips of 10 emerge from it to be stacked up in the tens bin. This photo shows the transition as the last strip is removed from the package. When the package is thus emptied, there remain 5 packages, 12 strips, and 4 balls. The paper-and-pencil form is not shown in this photo.

teacher's "occasionally," saying, "Oh yes, I use them all the time. I type all my homework on the computer." A girl in the group ribs him, saying "It probably has to re-spell everything you type." Smiling, the boy flashes back, "How can you say that?! I never make more than fifty or sixty mistakes in a sentence," and the whole group laughs. You are then told that the writing group rarely uses the computer during class time, but the students are encouraged to type all of their written compositions into a computer so that they can easily edit them when they discover improvements in wording, spelling, structure, or content. Because such changes never require the retyping of unchanged parts, the youngsters usually purge all the superficial errors of "mechanics" of their own accord. Further, many are encouraged in their writing by the fact that the finished work always looks so neat.

How long has this special learning center existed?, you ask. Not long, of course, the teacher explains, and the school is still feeling its way. Both money and technical support were sought from outside sources. Some computer programs were written by local high school students; some were the result of research projects at a local university. But in-school resources are developing, too. Some of the teachers in the school have really dived into programming themselves. The relationship with the community has been very exciting. After seeing Tammy learn to draw at the school, a local charitable organization raised the funds to give Tammy her own personal computer to use at home, and that generated more community interest and brought in more offers of programming help. Community members who volunteered time to scout out appropriate charitable foundations and other sources of potential funding made a particularly valuable contribution to the school's effort.

As you were reading the previous account, did you wonder if the school could be real? The answer is as with most of the vignettes in this book: yes and no. Every one of the activities described in this scene is or has been in practice, somewhere. There exist a variety of environments serving handicapped and non-handicapped students of different ages and academic levels together. The mix of computer-aided and non-computer-using activities in the same work areas is drawn from real schools. Tammy is real (though the name is changed); she drew for the first time in her life at age 15, and she is getting her own computer to use at home, compliments of a Rotary Club. Alas, what is not yet real is a single public school setting where all of this can be seen together. It is just around the corner, though—a very near corner, so near that we must think today about what we would and would not like such a learning center to be.

* * * * *

You are visiting a large research computer facility in a major university. Eden, an administrator, teacher, and scientist in this facility, brings you to her office within the sprawling computer room to show you a program that allows users of the computer to communicate with one another in different parts of the building while they work. There are several other people in the large computer room, all busily engaged at their own terminals.

The computer program that Eden shows you has two modes of operation. One mode is a little like holding a telephone conversation through typing rather than speech. A person using the computer calls up another person (also currently using the computer) by telling the computer to link to that person. If the other person accepts the call, then, until the link is broken, anything that either person types appears on both of their terminals. Thus, as the first person types a message on his or her own terminal, the same message appears on the other's terminal. The other can respond simply by typing back. A second mode is more like sending mail. A message can be composed in advance, and then sent, all at once, in a flash to the other person. In this case, the recipient of the message need not even be using the computer at the time. If he or she is not present, or is not accepting messages, the computer holds the message until the next time that person **logs in.** If the person is present, and is accepting messages, the message will appear immediately.

As if it had been planned, the screen flashes while Eden is explaining the system, and a message comes through. It is a request for help in using a program.

> From: RGB at CAIL
> Date: 2 Jan 1984 1626-EST
> To: EDENJ at CAIL
> Hi, again! Still more difficulties with SYMALG, and a few more questions: . . . .

Eden points out the different parts of the message. First there is the sender's identification, in this case just the initials RGB and the computer from which he is sending the message. Oh, yes, she explains, this communication facility connects computers all over the United States, and in a few foreign countries! Then there is the date and time of the communication. Then a note indicating to whom the message was sent. In this case, only Eden was a recipient. Finally, there is the message itself.

Eden mentions that she has often received messages from RGB, who happens to work in the same building, as indicated by the code "CAIL", but she has not yet met him. She types a quick reply and sends it off. A moment later, a thank-you note arrives with a continuation of the question. She replies again. When a third query follows shortly she suggests that it sounds like they have more talking to do than is really practical in typed messages. Wouldn't it be more efficient for them just to get together face to face and discuss it, she asks, especially since he must be working fairly nearby in the building. A moment later her screen flashes again with a note from RGB apologizing for not introducing himself in person, but adding that it would not be easier for them to talk face to face, even though, at the moment, he happens to be working in the same room! "I am," he reports, "deaf." You look up, and a few terminals away, a smiling young man in his early 30s waves a greeting.

Later, via computer conversations, RGB told of the much greater ease of "getting to know someone" through computer mail than when he and others had to handle the awkwardness of their first meeting face to face. In fact, he had corresponded for years and had formed close friendships with several people at a computer laboratory 3000 miles away, and they never knew he was deaf until they finally met at a conference. By then, the awkwardness of face-to-face communication had been overcome by the already-established bonds of common interests and shared stories.

RGB had lost his hearing when he was about six. He grew up entirely with hearing people and had no more contact with deaf children or adults than the average hearing child has. He learned his part well. His speech, though it lost some of its former distinctness, remained intelligible. He learned to speechread remarkably well, considering how extremely impoverished and unreliable a means it is. And, having grown up this way, he learned to feel that kind of communication to be normal. But in college, he discovered other means that were easier. The computer was the first of these, and that experience in large part shaped his career.

He also discovered sign language. He and a community of hearing friends learned the language at first on a lark, but later they began to feel that it was a genuine help in communicating, even though they had already developed their friendship with less fluent means such as speechreading, writing, and typing. Though speechreading had seemed adequate for communicating essential work and school information, social chatter was seldom repeated or redirected for RGB's benefit. Similarly, while the computer allowed for social chatter and fluent conversations in the office, it didn't help out over dinners or at parties where the face-to-face contact was an essential part of the fun. Signing made all that available to him, and made interaction with a group of people much more satisfying.

RGB is now a top-flight and highly respected computer scientist who has worked on projects for universities, government, and industry.

* * * * *

Intrigued by what you've seen in a school and in a workplace, you decide to see what possibilities computers offer to handicapped children at home. Tammy's home machine is not yet installed, but Ellis Central puts you in touch with a young writer, Lucien, who has been collecting material for a biography of four cerebral palsied children and has been volunteering some time in the writing lab in their school. One of these children, Eric, has a remarkable computer that goes with him wherever he goes.

Until the end of the last academic year, Lucien had been teaching in a private school for gifted children about a hundred miles away, but during the summer he and his wife moved here so that she could accept a faculty appointment at the university. Now he concentrates on his writing.

Recalling your visits with Eden and to the school, and excited by the little the school had been able to tell you about the CP boy with the computer, you ask Lucien if there is a way for you to visit that child at home.

There isn't.

Perhaps the most pervasive truth that Lucien is learning in his interviews with the parents of these children, he explains, is that their lives are filled with a near constant stream of professionals—teachers, therapists, counsellors, advocates, physicians, social workers, even journalists from the local papers. Unusual circumstances, such as Eric's computer, attract even more attention—photographers, wider press coverage, and requests to make public appearances. Eric's family also lives with frequent visits of the engineers and university students who are helping to design Eric's device.

Of course, there really is no stopping this publicity. Clearly, Eric and his family benefit directly from some of it. They also care very much about helping to show others what can be done for children like Eric, and so they accept the reporters and public appearances. But they also need to reserve some space for privacy that is so important to a normal family life. Lucien has, by this time, become something of an auxilliary member of the family, and suggests that instead of a visit, you might read an excerpt from the diary he is keeping in preparation for writing the biography of Eric and the other children.

SUNDAY, MAY 15, 1983—
OUR FIRST MEETING

Eric's parents and I sat at the kitchen table, talking. Lisa, almost ten and very self-sufficient, set up a pot of water to boil and asked if we'd like tea. Eric, who had just turned seven this week, wasn't home yet. Ever since he was five, he has spent one afternoon a week with his friends David and Miriam, a young couple who met him as part of a kind of big-sister/brother program for handicapped kids. Today's outing was a special one in honor of his birthday—dinner out at a Chinese restaurant and a concert at Symphony Hall. The family is musical, and Eric has been to concerts with them before, but being out just himself with friends is different.

I asked about the big-sister/brother program and was told that an advocacy group had set it up as much to support the families of handicapped children as to provide variety for the children themselves. Eric craves attention and social contact just as any child might, but because of his severe motoric impairment, he needs even more time with others simply because there are fewer activities he can engage in entirely alone. The tremendous amount of work involved in raising a child like Eric is not affected much by the friends program, but the little bit of time off is an important relief anyway.

Lisa brought the hot water and tea bags and went to get some cocoa for herself. She knew I was interested in Eric's computer and told me that she, too, was learning about computers in school. I asked her which computer she found more interesting, the school's or Eric's. She said that all the other kids thought Eric's computer was "wicked neat" but that she liked the one at school better because she could program it. She loved the computer art and music the best, and pointed to the refrigerator where several of her computer-generated pictures hung. She was sorry that she couldn't show me the newest picture she had made of three butterflies flying around a light, but I would have to come to school, she explained excitedly, to see her picture on the screen where the butterflies really moved just like real butterflies.

Outside, there were sounds of car doors opening and closing. Lisa announced that Eric must be home and fairly bolted toward Eric's bedroom to show me Eric's wheelchair-mounted computer before he got in, but her father said that he thought Eric would prefer to show it himself. It didn't quite make sense to me. Why would Eric ever be without his wheelchair and communicator? Wasn't portability its special feature?

It took a while before Eric showed up at the front door. He had to be taken out of his special handicapped-child's car seat, his fold-up travelling wheelchair had to be unpacked, and he and his belongings had to be gathered up before David and Miriam could wheel him in. Lisa, apparently sensing that her part of the show would soon be over, hurried to tell me that Eric's big chair could not fit in a car and could only be transported by van.

At the door, Eric greeted us with a wide grin and a lot of excited arm-waving, chirping, and kicking. Miriam began to say that he loved the concert, but Eric turned half-way around, one arm flying out in front of him, and grunted loudly. Miriam said no more and as she began to wheel Eric to his room, Eric stared hard at David. David promised not to tell any more about the day until Eric was back.

The sounds coming back down the hall were different. Miriam returned to the kitchen without Eric, and I heard none of the creaky sounds of the stroller-like fold-up chair. This time, there were the soft motor sounds of Eric's electric wheelchair. The soprano voice of a child of perhaps 12 or 13 called out "It sock pearl man played shy cough ski violin concerto." In fact, it sounded more like a few 12 year olds speaking, as the voice changed a bit from word to word. I must have looked as bewildered by the strange utterance as I was feeling. Probably more for my benefit than for Eric's, the parents repeated a lot of the words: "Itzhak Perlman, eh? Oh, and Tchaikovsky's violin concerto is one of your favorites, too, How was it?" To me they explained that Eric's communicator has over 3000 whole words in it, many recorded just to meet Eric's interests, but that words like "Tchaikovsky" had to be assembled from other words or the nearly 1000 word-parts that the communicator contained. As for the concerto, Eric often asked them to play it on their stereo. Eric appeared at the door in a most remarkable wheelchair, and the voice said "It sock had crutches."

Having prepared myself to see computers hanging all over Eric, I was most surprised by what seemed to be the total absence of anything that looked like a computer. For that matter, even the chrome-and-black look of hospital technology was lacking. Eric's chair was a beautiful piece of furniture, attractively upholstered, and form-fitting. It could drive forward, backward or sideways, and could swivel on its axis, allowing Eric to face in any direction, and to drive anywhere the chair could fit. At a convenient height above Eric's lap was a desk top with a small TV attached securely to it. Eric had an awkward grasp on an object which looked a bit like a very small clothes iron, and was "ironing" apparently patternlessly on his desk. The voice on the TV said "hello." Lisa introduced me. She also explained that the computer somehow knows where Eric is holding the iron and translates his ironing into commands to the computer.

I asked Eric if he enjoyed his outing, and he said yes. Eric fairly flew out of his chair as he turned his head and grinned at David, and then he got back to his ironing. There was more conversation at the table while Eric was busy "writing" on his computer tablet, but I was so riveted to Eric and his labors that I remember little that anyone else said. Eric finished composing his comment and "said" that David fed him with chop sticks at the Chinese restaurant and that he wanted to do that again!

We spent about an hour more before Eric had to go to bed. I learned that Lisa played the violin quite well, that Miriam and David were expecting a baby, and that Eric wanted to be in on everything. He, too, was eagerly awaiting this baby, and felt as though it would be part of his own family. "If it's a boy, I would have a brother and a sister!" Lisa had taught Eric how to play two short snatches of melody from the violin concerto on his computer's music synthesizer. Eric ironed some more and gave us a brief concert. Both selections were very simple— one mostly just a descending scale repeated twice—but the music was quite recognizable, and this 7-year-old severely handicapped boy was playing it!

As in the story of Ellis Central School, the technology described in Lucien's chapter is not a fantasy. The attractive, omnidirectional, computer-controlled wheelchair exists. In fact, even wheelchairs that respond to their owners' spoken commands are becoming more widely known and used. The ability to detect meaning from the seemingly patternless movements of Eric's hand—the same technology that allowed Tammy to draw the letters of her name—has been demonstrated. So are the speech production and music synthesis. The social picture of Eric and his family—their own nature, the fishbowl that they live in, even the attention of a professional writer—was drawn from Erics we've known.

Also, as in the school story, the fictional element is the creation of a composite. We know of no Eric whose communication system incorporates all of these features. However, the consequences of the lack of some particular aid are much more severe for Eric's personal life than they are for a school; for Eric, each aid is freedom. Furthermore, the Ellis school fantasy is just around the corner, but Eric's rich life seems not yet to be. If social policy has been less than magnanimous to education and educational institutions, it is downright unfriendly to the Erics of the world. Whether we think only about governmental priorities or take a broader sociological perspective, people who lack are pushed to the fringes of society. As long as Eric's most striking attribute is a lack rather than a strength, he is being hurt. Eric is not children-in-general. He cannot wait. As time goes by, he grows older, and his losses grow, too. Opportunities to educate him are passed by, and his choices diminish.

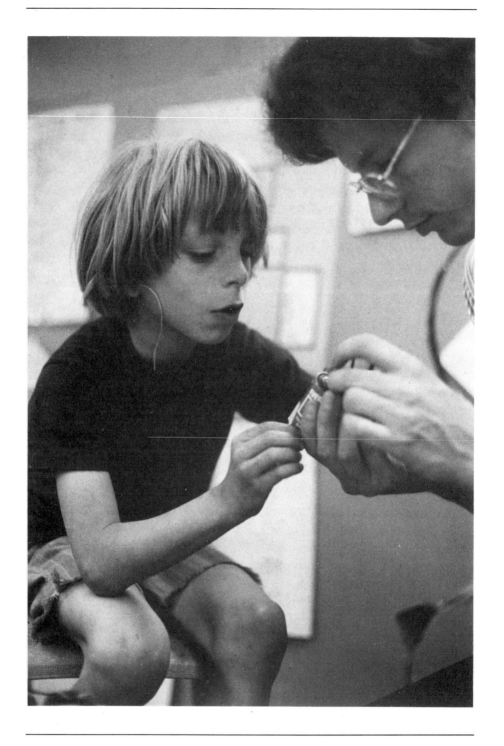

Education and Special Needs 2

Like Eric's family and Lucien, we, too, want to spread the word about what has been accomplished. We also want to raise interest in what can be. Yet the task contains a paradox.

Computers, like blackboards, Cuisenaire rods, and cameras, are among many tools available to educators. None of these tools would be used alone in a classroom, and we would not speak of "camera-based education." Yet the newness of computer technology (and some of its characteristics) allows us to use terms like "computer-based education" quite casually. Nobody writes a book entitled *Blackboards, Education and Special Needs* and yet you are now reading *Computers, Education and Special Needs*.

This isolated treatment of computers is something of an artifact of the times. Computers are new and blackboards are not, so a book is written. More important, computers may have a particularly great impact on education because they are a powerful information technology. Information is the substance of knowledge, and learning to manipulate knowledge is much of what education is about. Computers are information manipulators, *par excellence*. So it is that a book entitled *Computers, Education and Special Needs* can come to be.

Yet, in a very foreseeable future, computers will be as familiar as any other tool. At that time—as it was before computers became a reasonable topic for a book on education—a book about good educational techniques for the special needs student will again not focus on one particular tool, but rather on the learning process in the context of the whole educational environment.

We are concerned about that future time, in part because we want to bring it about. And, though we are very excited about the educational value of the computer-tool, we do not want the computer to be any more focal in the classroom than a blackboard should be. Good *teaching* is done by good *teachers*. Tools are merely tools. So, though the date constrains us to focus our attention on the new technology, we have endeavored in this book to show it in an integrated perspective.

The rest of the message, then, is about good education in a special needs setting. But how might that differ from good education anywhere else?

It is true that the students represent a select group, but it is certainly not a homogeneous one. Specialized teaching techniques and knowledge may be needed to deal most effectively with mental retardation, blindness, cerebral palsy, deafness, or learning disability, but it is no more informative to classify these techniques together under the heading "special education" than it is to classify methods in mathematics, language arts, and physical education together under the heading "regular education." They are all specialized knowledge, and the classification "special education" adds little except a stereotype to that knowledge.

Law has mandated the development of individual educational plans (IEPs) for any student designated as "special," and has not made similar requirements for other students. Yet, that does not represent a philosophical difference between the two educational practices. Good educational practice tries, within the limits of practicality, to recognize individual differences, set appropriate goals, and maximize the realization of individual potential for all students. The fact that law requires us to formalize this for some and not for others—be it blessing or curse—is not a statement of philosophy or psychology, but a recognition that we cannot get by without flexibility and individualization in the special classroom and yet can at least pretend to do so in the mainstream.

Of course, this very fact says that the classification "special" is important. Though it may be only a matter of public image and politics, it has implications for funding, legal protection, self-image, social acceptance, treatment in schools, and more.

All of this is not to say that there are no pedagogical issues. There is a lot to say, for example, about communication, accessibility of content, and motivation, but these suggest special techniques, not special philosophies. Good teaching remains good teaching.

LEARNING AND EDUCATION

Learning is a life-long process that begins at birth, and can occur anywhere: home, neighborhood, school, pre-vocational and vocational training programs, colleges and universities, in the workplace, and in the great outdoors. People are constantly learning.

Education, on the other hand, usually refers to learning that is mediated by an other; in school this person is generally the teacher. While philosophies of education vary widely in the precise roles they ascribe to the teacher and to the learners, most agree that for learning to be effective, learners should be motivated, they must have access to the "stuff" of learning, and they need some form of communication as they engage in educative activities.

Education in formal settings such as schools has traditionally involved the teacher in providing this motivation, access, and communication for the learners. Although this tradition may be convenient, perhaps even efficient, it carries with it certain psychological liabilities which, in the case of special needs learners, may even be counter-educative. For some learners, autonomy may be the key to cognitive and emotional growth while overdependence on teacher-centered education may block opportunities for becoming more autonomous.

Yet, how else is mass education to be carried out, particularly for children with a wide range of physical, psychological, and social handicaps? There is no simple answer, but we suggest that it may be possible for computers to be used in ways that simultaneously provide motivation, access, and communication on the one hand and promote autonomy on the other. That is the subject of this book.

SPECIAL NEEDS IN EDUCATION

The term "special needs" is a misnomer. Students with handicaps have the same needs as any other students—opportunities to learn, to play, to form relationships, to exercise autonomy, and so on. What is different is the diminished access handicapped students have to these opportunities, and what is special are the techniques that must be used to meet their needs. In fact, when a technique overcomes the handicap in a full and socially acceptable way, as is the case with eyeglasses for myopia, terms like "**handicap**," "**disability**" and "**impairment**" are no longer applied. "Special needs" then is really a socially determined shorthand for "need for special adaptations if *normal* needs are to be met."

One of the most important tasks of the special needs teacher is to recognize what adaptations are needed and to provide them, if at all possible. This is also one of our most frustrating tasks. Sometimes the conditions blocking learning are so great—multiple and profound handicaps—that only extraordinary interventions would appear to be able to meet even the most minimal needs of the learner. Worse, these interventions frequently are unavailable due to high cost and different social priorities. Other times, the conditions blocking learning are so subtle—certain learning disabilities or behavioral disorders—that they masquerade as other problems and thus defy intervention.

Nevertheless, providing appropriate adaptations to facilitate learning is the job of the special needs teacher. Let us consider several case studies, asking with each one what kind of intervention might provide the student with opportunities for learning.

MOTIVATION

Motivation is a key to learning for all people. If there is no felt need to learn something, the chances of learning it are slight. What are and what should be the sources and types of motivation for learning have been the subject of debate among philosophers for thousands of years and among psychologists since the founding of psychology. It would not be fruitful to rehash those debates in this book. Suffice it to say that motivation, whether externally applied or internal in origin, whether a physical need or a psychological state of mind, is a prerequisite for most learning. Among special needs students, motivation is a particularly acute problem since much of their life experience would suggest that academic investment is unlikely to improve their lives in the ways they would wish. For such students, motivation includes far more than the will to learn; it involves the will to be. Thus, a sense of efficacy and autonomy are key elements in the motivation of special needs children.

Emotional Difficulties

Lisa, who is now nine, lived with her biological parents until she was four. When it became apparent that she was being abused by an out-of-control parent, the court stepped in. The parent would not get help and chose instead to put Lisa up for adoption. She was adopted by a family who for unknown reasons decided they could not keep her. After only three weeks with her second family, she was placed with a third family where she has been lovingly cared for and nurtured ever since. Nonetheless, she has such extreme fear of displeasing or being rejected by another adult that her anxiety nearly paralyzes her in most learning situations.

What kind of strategies and materials could a teacher use with Lisa to help her overcome her anxiety and fear of rejection and be ready to take risks and learn?

Learning Disabilities

Learning disabilities can affect a student's motivation to begin or stick with a learning task. Mark is an example of such a student. He is a bright 13-year-old who has a learning problem of undiagnosed origin. He has great difficulty focusing in on whatever he is doing. He does not seem to be able to shut out what for the other students would be extraneous sights and sounds. If an adult is not sitting beside him to draw his attention back to the task at hand, Mark is likely to wander off, attracted by something which is either externally or internally more stimulating.

Mark's mother says that he is the same way at home. He leaves games, projects, snacks unfinished. There is only one activity which holds his attention for more than two or three minutes—playing video games. Mark's mother is convinced that the roof could cave in around him and Mark would continue to play undisturbed!

What special adjustments could a teacher make to help Mark have as much interest in a classroom learning activity as he does in his home video games?

The Effects of Being Exceptional

Students who cannot or do not learn in the usual ways are further hampered by their experiences in our various institutions for mass education. Objectively, these students experience frustration and failure in schools that are ill-equipped to meet their needs. But perhaps more important, subjectively they experience the debilitating effects of being set apart, locked into a special social and educational environment. In particular, these exceptional learners often encounter responses to them that are denigrating, defeating, destructive—responses that range from fear to pity. Primary emphasis in our society is placed on what these students can *not* do. Moreover, when teachers, parents, friends, and public feel sorry for someone who is retarded, blind, deaf, uses crutches or a wheelchair to move about, they feel they should do *for* that person. Or often people do things for handicapped individuals because it is easier than devising ways for the learners to do for themselves. People have many of the same reactions, though perhaps in more subtle forms, to the "invisible" handicaps of dyslexia, math phobia, etc. Under these conditions and the weight of pejorative attitudes, developing a sense of efficacy is a difficult task.

Although George, a bright but obviously immature, 17-year-old with cerebral palsy can get around school quite well on crutches, his parents prefer him to use a wheelchair for fear of him falling and hurting his knees on which he has had numerous operations. George wants to go to college, but his parents have discouraged him because there would be no one to take care of him there. George's older brother and sister were both popular, athletic, college prep students, but George hangs out with much younger students, mostly social misfits and boys of marginal intelligence. George had done rather well academically up through tenth grade, but soon after his guidance counselor and parents had discussed college he began to lose interest in his classes, started staying home from school frequently with "sore knees," ceased doing his homework altogether, and withdrew into an angry shell.

What can George's teachers and guidance counselor do to help him regain the motivation that will allow him to feel autonomous, to make a break from the well-meaning, but debilitating influence of his parents?

ACCESS TO INFORMATION

Handicaps—often even the subtle ones—can limit access to learning via conventional means so profoundly that it is attainable only through great struggle and superhuman determination. Compounding the difficulties that are the direct result of the handicapping condition are artificial barriers—overprotections and underexpectations—resulting from attempts by others, and sometimes by the individual himself or herself, to cope emotionally with the handicap.

Self-Protection as a Barrier to Information

Margaret Riel studied the problem-solving strategies of elementary school children who were two to four years behind their age level in language development. Though conventional testing of these children suggested a general cognitive deficit, Dr. Riel's studies presented the alternative explanation that these children, in an attempt to reduce their frustration and "to pass as normal," developed successful ways of avoiding the situations in which their failures were most apparent. By choosing unrealistically difficult tasks, they could look brave and attribute any failures they might have to the task and not to themselves. By claiming to know already how to play a game, they could avoid the frustrating instructional interaction with others and learn, if necessary and possible, "on the fly." However, while these techniques succeeded in covering up some of the students' weaknesses, they did so at great informational cost which showed up as slow learning. Acting on this observation, Dr. Riel modified the classroom environment so as to remove the self-made barriers to information. When this was done, these students showed themselves often as capable of achievement as were their language-normal classmates. Without that restructuring of the learning environment, these children's avoidance strategies limited their learning opportunities well beyond the "natural" limitations that resulted directly from their reading or language impairment.

"Natural" Barriers to Information

The word "natural" is misleading. Self-protection is as natural as conditions such as the reading impairment, deafness, learning disability, or motoric dysfunction that might lead one to it. Still, sensory impairments including deafness and blindness, as well as perceptual, language, or neuromuscular disorders are often primary, in at least the sense of "first," giving rise to others.

Most individuals who are **prelingually deaf,** deaf before they are three, are not as fluent in the spoken language of their environment as those who hear that language spoken to them each day of their lives. Further, though these individuals may have no visual or cognitive deficit, their reduced knowledge of the language of the common culture hampers their reading of that language. As most conventional learning situations depend on reading proficiency normal for the student's age, students beyond the earliest grades are severely restricted in all of their learning by a weakness that may be specific to reading alone.

Other students fail to learn the surrounding spoken language, not because they cannot hear it, but because they cannot structure the sounds that they hear into a coherent message. Again, though there may be no visual or other cognitive impairment, reading is affected because the language is foreign to them.

Impairments of visual sensation and perception, while not diminishing the acquisition of language, similarly reduce access to its printed forms. Learners with perceptual difficulties may see print well enough but not understand it because they are unable to organize the printed symbols into meaningful units. For others, sight is the limiting factor. Even with the use of corrective or magnifying devices, print materials including writing on a blackboard, textbooks, library books, and self-made notes are not accessible.

For some of these learners, hearing is a reasonable alternative, but having to depend always on another person to read, or on the availability of recorded materials consumes time, reduces autonomy and independence, and often limits the scope of one's endeavors. Dennie, a **blind** doctoral student, faces the frustration of scanning large volumes of text to find the few paragraphs that are of real importance, a task that, at times, becomes impractical as she is not reading for herself. Though Braille increases the independence with which some students can read, it does not solve most of the availability issues, and is not very helpful to Dennie in her research.

The success of many deaf, blind, or dyslexic students even in advanced graduate studies certainly attests to the possibilities, but probably also reflects determination, opportunity, and good fortune that have skipped over many other students with equal intellectual ability. The same applies to other nonreaders and poor readers, though the causes and other implications of the deficiency may differ markedly.

Physical access is the barrier most publicized by our society. But reserved parking spaces, ramps, and elevators do not solve all such physical access problems. Cheri, who, as a result of **spina bifida** is paralyzed from the waist down and confined to a wheelchair, will enter high-school chemistry this year. Her teacher is wondering what sense there is in teaching the care and use of lab equipment to Cheri if she can't work at the lab table. The table doesn't accommodate her chair; she cannot get close enough or high enough. Joe, another student slated to be in the same laboratory class, has the physical access, but has severe motoric impairments affecting his hand and arm coordination that make it appear unsafe for him to handle the equipment and frequently caustic materials in the chemistry laboratory.

Inappropriate Expectation as a Barrier to Information

While some of the problems that these students face would be alleviated by some form of communication enhancement (a reading prosthesis, improved language training, a breakthrough in the teaching of reading, etc.), or the removal of architectural barriers, much of their handicap is essentially social in origin. Expectation influences outcome. The kind of children with whom Margaret Riel worked appear to most people as generally slow learners and poor students, with the result that less is offered to them in the way of learning opportunities, and even less is expected of them.

Joe, who can raise a tall overfull glass of Coke to his lips without spilling a drop, might well be able to handle without risk more in the chemistry laboratory than people expect. Cheri succeeded well in biology laboratory in which many of the experiments could be performed on a board at lap level. Yet many children who are deemed "exceptional" are denied even those experiences that they are perfectly capable of handling. Children who cannot see, or cannot hear, or are slow to learn, or who cannot read, or who must sit in their wheelchair observing while others perform the manipulations may be seen as "not able to appreciate" the experience, or it may be felt "too difficult to explain to them." The result is an impoverishment not only of what they glean from their experiences, but even in what is offered them. And the effect persists into adulthood.

The popular misconception that deaf people or cerebral palsied people cannot drive, for example, often delays the time when they get to discover that they can.

Some of the experiential deficits are quite early. Sally is five. She has had spastic cerebral palsy since birth and, as a result, the voluntary movements in her hands, arms, and legs have always been restricted in accuracy, speed, and fluency. Her curtailed mobility and hand coordination have prevented her from having all sorts of typical life experiences. She goes from home to school, transported in a wheelchair van, and occasionally down two flights by elevator to her grandfather's apartment, but that's just about all. The family does not own a van, and Sally's mother says that Sally's braces and wheelchair are just too much to cope with after a full day's shift at the restaurant. Sally has been to the corner grocery a couple of times, but she has never since infancy been to any other store, museum, park, zoo, post office, bank, restaurant, movie. Since she was two, Sally has not ridden in anything but the school van. She has never played in sand, dirt, or snow. She does watch television, but that does not begin to give her the kind of background she needs for her learning at school.

Alternatives and adaptations need to be found to meet the needs of students such as Sally, not only for access to the content of a particular learning activity in school, but also for access to the experiences that are prerequisites for other later learning. The cycle is complex. To get parents and teachers to offer more varied experiences to children like Sally, we must in part raise *their* expectations. Yet, it is cruel to do so unless there are reasonable prospects of success. Where do we start?

COMMUNICATION

Learning of any kind involves communication between the learner and the learning environment, whether the latter contains a teacher or not. Students take in information or data from their environment, process it internally, express it in some form, and receive feedback on that expression from the environment. The feedback may come in the form of teacher approbation or simply from some facet of the environment responding to the learner's output (e.g., a balloon popping when too much air is blown into it).

Communication of all kinds requires reliable means of making output and receiving input. Communication in educational settings is particularly demanding of input and output channels because of the large volume of information that must be communicated and because of the need to constantly cycle through input-output-feedback loops if learning is to be effective.

Quite obviously, students who have difficulty hearing or seeing, students who have difficulty speaking because they have never heard language, or because their speech is unintelligible, and students whose language disability renders their communication incomprehensible will need adaptations and adjustments to participate successfully in the educational process. Thus, as with the need for access, if one sensory channel is not available, can another be used for communication in such a way that it will provide an adequate substitution?

Alison has an auditory impairment which severely restricts her hearing in the speech range. As a result, much conversation is lost to her. The nature of her hearing loss is such that amplification of what she hears by way of a hearing aid or other device does not help much. How can Ali communicate with her teachers? How will she be able to ask questions about what she does not understand? How will teachers be able to respond to her questions? Some alternative means of communication is needed—one that will provide the same rate of information that occurs in conversation, that will be mobile, and that will always be readily available to Ali and her teachers.

Similarly, students who cannot communicate through writing need an alternative form for expressing themselves and for using what they have learned, which would be as flexible for the student and as accessible to a teacher as paper and pencil.

How many times have teachers wished they could translate information, theirs or their students', from one medium to another—make visual displays of auditory information or auditory displays of visual information; create visual representations of concrete objects that could be acted upon, turned around, and manipulated; change text into speech, Braille, or tactile displays?

THE TASKS OF A TEACHER

Above all else and regardless of educational philosophy, the central task of the teacher is to facilitate student learning. Whether the environment is teacher-centered, subject-centered, or child-centered; whether the approach is prescriptive or exploratory; whether the techniques are behavioral or humanistic, the job of teacher is fundamentally the same. Moreover, to promote student learning all teachers, formally or informally, constantly assess student learning, maintain a learning environment (or curriculum), and manage more or less information about the students and their learning. Consider two sharply contrasting cases and note their fundamental similarity.

Susan Cohn and Bill Jones are teachers in an integrated day, open classroom, containing thirty-five boys and girls, aged nine through eleven, including five "mainstreamed" students with a range of special needs. The room is divided into learning areas: a science-kitchen lab, a computer corner, an arts and crafts area, a discussion circle, a quiet "room," a library-media center, and the teachers' space. Students move from area to area on their own. Susan and Bill spend varying amounts of time in each student area, as well as in their own space. Usually they are talking with students, although sometimes they confer with each other about something they've observed or a new idea for materials or activities to be added to the rich environment they've already created. The only requirements students have are to file a learning plan at the beginning of the day in which they write down the activities they think they'll do that day, as specifically as they can, and to reread the plan at the end of the day, indicating on it what they accomplished from their original plan, what else they did, and what, if anything, they wished they had done. Sometimes students, of their own accord, add some evaluative comments, such as: "I was a goof-off today, don't know why but I hope I feel more like schoolwork tomorrow" or "Boy, was that fun. I'm going to do it again Friday after I finish my math worksheets."

The teachers' day often doesn't end until five at night. They generally spend some time after school decompressing with a cup of coffee or tea, sharing observations about the day. Then they divide the student learning plans between them, read them, and make notes in a separate folder about the student's learning, based on the students' own accounts and the teachers' observations. Once each quarter, the teachers meet with each student to discuss their learning experiences, and once each semester one or both parents join this discussion. At these conferences, reference is made to the students' folders, containing their learning plans and any other output they've chosen to include, as well as to the teachers' folders for the students.

* * * * *

Sam Parker is the principal at Barrows Elementary School, a demonstration-lab school for the University Special Education Department. All of Sam's teachers are graduates of that program and Sam himself is a Clinical Professor in it. The school is run entirely according to behavior modification techniques for both the normal students and the special education students, who constitute 30 percent of the student body. In the early grades, a highly structured program developed at another university is used for reading and language acquisition. Sam selected this approach and himself trained the first- and second-grade teachers in its use. Other curricular approaches were determined by a committee, chaired by Sam, consisting of one master teacher from each grade, and the various specialists in the school. The entire K–6 curriculum is coordinated to minimize redundancy and yet reinforce skills learned previously. Testing is done daily as students complete each learning module. Results are scored and stored on the school's computer in a system designed by Sam and the computer specialist for the district. At the end of each quarter, more extensive testing is done to assess retention of learning, and at the beginning and end of each year testing is done over the whole range of the curriculum. Also, at various times during the year, IQ tests and other nationally normed achievement tests are administered to monitor progress of individuals and the overall program. All test results, from daily competency assessments to nationally normed standardized tests, are shared with students, their teachers, and parents.

In both of these extremely contrasting cases, teachers facilitate learning by providing learning environments, assessing student learning, and managing information about that learning. In both, it is necessary to make provision for those students whose physical, cognitive, social, or emotional characteristics hamper access to the learning environment, block communication with that environment, and dampen motivation to participate in the learning process. The teachers' task in both cases is a formidable one as it relates to many of the students.

POSSIBLE ROLES FOR THE COMPUTER

The advent of relatively low-cost, portable, yet powerful computers offers special educators an opportunity to vastly expand their repertoire of interventions and adaptations. This is certainly not to suggest that computers be used for all interventions. There are cases in which other techniques would be more appropriate, less expensive, more direct. Recall George whose motivation was sharply reduced by his over-protective parents' scepticism about him going to college. There is little about George's problem that would suggest a role for a computer.

There are even cases in which computer use could be counter-productive. Take the case of Sally who only has first hand access to home, school, her grandfather's apartment, and occasionally the corner grocery store. These extremely limited life experiences make Sally ill-prepared for later learning. What she needs foremost is to spend time out in the world with other people, observing natural and human-constructed phenomena, developing a range of interpersonal relationships, in short having many of the life experiences any 5-year-old would have. It is doubtful that a computer would greatly assist in this process and, in fact, were Sally to spend a great deal of time with a computer at home instead of "getting out," it could even worsen her situation.

Computer as Stimulus

On the other hand, there are countless situations where computers could significantly contribute to the student's learning situation.

Mark, whose attention to tasks was practically nil except if the task were to outwit threatening monsters on his video game set, is an obvious example. As it happens, Mark is a student in the sixth grade at Sam Parker's Barrows Elementary School. He was transferred there after the teachers in his old school informed his parents that they simply could find no way to get him to attend to lessons.

During his first week at Barrows, Mark underwent a battery of individually administered tests. Because of the exceptional skills of the testing specialist at Barrows, for the first time in his school life, Mark was successfully tested. The results confirmed the impression all had had that Mark was bright, although they provided little insight into the cause of his poor attention span.

After interviewing Mark's former teachers, his parents, and him, the learning disabilities specialist decided, in keeping with the school's philosophy, to create for Mark a program that would systematically increase the duration of tasks he was to perform, beginning at first with very short ones. The program utilized a point system in which points accumulated could be "cashed" in for a variety of rewards and privileges. The longer Mark attended to a task, the more points he accumulated.

One of the first rewards the specialist provided Mark was the opportunity to "play" an arcade-like educational computer game that was part of the *Arcademics** series. Much to the specialist's amazement, Mark played this game with full attention until she interrupted him with the news that it was time to go to lunch. The specialist had been told about Mark's fascination with video games, but because she suspected organic brain damage to be the cause of the attention span problem, she had hypothesized that he wasn't really playing these games according to the rules, but merely flitting from one special effect (sound, color, design) to another.

Thus, she was totally taken aback when she saw that not only was Mark playing the Arcademics *Meteor Multiplication* game by the rules, but that he was playing it successfully at a very high skill level. The specialist discussed the situation with Sam Parker who suggested that the cause of Mark's lack of attention might include an extreme need for stimulation. "In other words," Sam said, "he gets bored very easily."

To test this hypothesis and some consequent interventions, Sam and the school curriculum specialist constructed a new program for Mark, this time utilizing a wide variety of manipulables, color photographs, slides, and short 8 mm filmloops, as well as the few educational computer games they could find. Mark had the time of his life and, after a week,

---

*This and all other commercially available software mentioned in this book are listed in the Resources Section along with the names and addresses of their publishers.

was attending to tasks with these colorful, concrete, interesting materials for large chunks of time.

It is interesting and important to note that although computers provided the initial clue to Mark's eventual "treatment," later, non-computerized interactive, stimulating materials were used side by side with similar materials on the computer. This theme—computers providing the initial clue, but not remaining the major emphasis in "treatment"—is one that is repeated throughout this book. Computers are a tool, though not the only one, to promote learning for students whose needs are not being met.

Computerized Access

Access has been the primary focus of legislation for the handicapped over the last several years: access to public transportation, access to a variety of public facilities, access to the "least restrictive environment" for education. However, for many, it will take more than legislation to assure access, so great are the limitations of their physical, psychological, or mental nature. A bright blind student such as Dennie might be assured entry to college, but once there how does she gain access to the educational resources that depend so much on sight? Physically handicapped students like Cheri and Joe, who this year are signed up for chemistry, may be mainstreamed into the regular academic program of their high school, but how can they engage in activities with which their physical disabilities seriously interfere?

Imaginative educators and engineers have begun to recognize the great possibilities of computers being used to compensate for the loss of normal means of access to educational and other resources. It is likely that over the next decade developments in robotics, in voice-activated computers, and in artificial intelligence will result in significant improvements in access for people whose nature blocks access even when institutional and architectural barriers have been removed. But it is not necessary to wait for these sophisticated technological breakthroughs to provide greater access. Already a wide variety of relatively simple devices exist which can transform today's reasonably unsophisticated computers into meaningfully effective prostheses. **Light pens, graphics tablets, touch sensitive screens, joy sticks, speech synthesizers, speech digitizers, voice recognition programs,** and a variety of interfaces to other input/output media are available

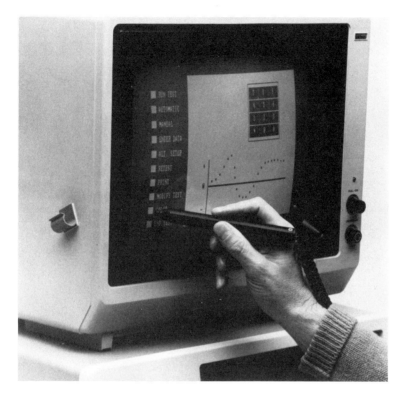

Fig. 2.1 A light pen being used on a VDT.

for the thoughtful educator to apply in any number of situations to overcome blocks to educational access.

This book is filled with examples of such **prostheses** being used, and sources for them may be found under Hardware in the Resources Section. However, two caveats are essential. First, the state of the art is such that the use of most peripheral devices for prosthetic purposes still requires a modicum of hardware sophistication on the part of those involved. But perhaps more important is the recognition that the physical device represents only a portion of the solution. Prosthetic use of the computer requires thoughtful design of an educational environment by the teacher. This can be a complicated and time-consuming process.

Cheri and Joe's chemistry teacher, Charles, was used to handling all sorts of equipment for lab experiments in his classes, but was insecure

Fig. 2.2 A child using a graphics tablet.

about his knowledge of physical handicaps and so was uncertain about how to modify the laboratory environment for Cheri and Joe.

Fortunately, Charles' school was located near a major university with a large and varied computer science department containing a research group specializing in computer applications for the handicapped. Charles and the special education coordinator for the district met with a member of the research team to discuss computer-prostheses for Charles' science laboratory.

After Charles described the nature of Cheri and Joe's difficulties in the lab, the special education coordinator inquired whether Charles had

Fig. 2.3 A touch sensitive screen being used.

Fig. 2.4 A joystick with a large print computer (Visualtek).

considered simply assigning Cheri and Joe to lab partners who could perform the physical manipulations while Cheri and Joe engaged in the mental operations needed for the experiments. Charles *had* considered this option, but felt it to be too limiting in many cases. He wanted Cheri and Joe to get a "feel" for laboratory work and to develop a capacity for testing their hypotheses spontaneously without needing a lab partner as an intermediary.

The solutions arrived at for Cheri's and Joe's difficulties illustrate the potential and the limitations of computer prosthetics and the need for open-minded problem solving that includes computer possibilities, but recognizes non-computerized solutions, as well. The solution for Cheri, for example, did not involve computers at all; it was decided simply to have some students in the vocational woodworking program build a lab station that could be placed at the appropriate height over the front of Cheri's wheelchair like a large serving tray. Once the idea emerged, Charles proved to be an imaginative designer, thinking of all sorts of ways to build in safety features to protect Cheri from spillage or other laboratory accidents. Still he recognized that there would be experiments in which the danger of working so close to chemicals would make the lab-partner approach more desirable.

For Joe, the solution was less satisfactory, although use of the computer may have contributed enough in unintended outcomes to compensate for its failure to totally meet Charles' goals for Joe. It was determined that Joe's erratic hand movements were simply too severe to allow him hands-on experience with chemicals and glassware. So the computer scientist proposed that Charles and he work together to create simple computer simulations of the experiments that Joe would ordinarily be doing in the lab. Then, Joe could "perform" the experiments on-screen while his classmates did the same experiments at the lab tables. Charles was excited by this idea, but pointed out the difficulties Joe would have with striking the right keys on the computer keyboard. The computer scientist was unconcerned about that problem. He assured Charles that he could easily create an alternative input device that Joe would be able to control.

Although Joe was never allowed to handle chemicals and equipment directly, the simulations created for him and his special input controller had the unforseen benefit of making Joe and his lab-station the center of attention in the lab and the object of not a little envy. Unlike the "real" experiments of his classmates which often went amiss due to human error, sloppy technique, and Murphy's Law, Joe's experiments always "worked." Moreover, Joe and his classmates quickly discovered that with Joe's simulated labs, they could easily experiment with the variables in the investigation to test different hypotheses. For many students in Joe's class, laboratory investigations became more than cookbook exercises for the first time. The speed with which Joe's simulation could "carry out experimental procedures" enabled students to explore phenomena in ways that would otherwise have taken far too long to investigate in the limited time allotted for high school lab work.

The materials developed by Charles and his collaborators thus benefitted all the students in his class, those with special needs and those without identified special needs. The key to their approach was to recognize where computers could make a difference and where other interventions might be more appropriate.

Joe and Cheri were fortunate. Their school had access to unusual resources at the university. Most students are not so fortunate. Take Dennie, for example. Her college and now graduate school lacked the human and financial resources to adequately address her need for more efficient access to written materials in textbooks, on the blackboard, and in the library. Although the existing reading machines for the blind might have helped her tremendously, they are not adequately accessible themselves, in part due to their cost. And, computer-aided educational materials for the blind are, as yet, quite limited.

Computers and Communication

Computer prostheses for the hearing impaired have undergone significant developments over the years, having begun even before the advent of the recent microcomputers. The computer itself, with its visual output, is a prosthetic for the deaf and, with the development of interactive programs, it has been an increasingly responsive learning environment for them. In the future, we can expect computers to play an even more significant role in the education of the hearing impaired, as developments in artificial intelligence make computers more and more human-like in their responsiveness, as portable speech recognition devices become more sophisticated so that the deaf will be able to "read" what people say, and as communications networking with computers enables the hearing impaired to communicate directly with others just as most people now communicate via telephone.

As an example of what is possible already, recall Ali, who has a significant hearing loss. Fortunately for her, she is a student in Susan Cohn and Bill Jones' open-classroom where she spends a large part of her day in the computer corner. Ali has learned to use the computer as her window to learning. She has devoured every piece of educational software in the classroom library, even those considered to be a grade or two above her. Moreover, she is constantly scanning catalogs and magazines for computer software that sounds interesting to her. These she asks Susan or Bill to try to get, at least on loan, for her to use.

Ali's parents have decided that she should not learn to sign, although many of her classmates have chosen spontaneously to do so, so her interpersonal communication is restricted to that which is directed to her personally. Susan and Bill never lecture so Ali doesn't miss out on information traditionally imparted in this way. However, group announcements have to be presented specially to Ali and she does miss out on much of the information transmitted by her teachers and fellow students in the myriad informal conversations and discussions that take place constantly in the classroom. Most conversation is not deemed worth repeating, and Ali simply misses it. But in the computer area, casual communication is easier. Other people working nearby can carry on a computer conversation with Ali by sharing the use of her keyboard. Although somewhat slow, these computer conversations have resulted in an impressive development of typing, spelling, and reading skills in Ali and many of her classmates.

Ali's dad, an engineer, is planning to buy a computer for Ali to hook up to their telephone at home and to provide the school with a **modem** so Ali can communicate with her class when she is sick at home. This is only the beginning for Ali. Bill and Susan hope eventually to network the computers in the classroom so that Ali can more easily carry on computer conversations with others.

While physical blocks to communication such as hearing or vision impairment are the most obvious, emotional communication blocks are just as problematic for learning. It is in this area that computers may be able to make perhaps their most surprising contribution.

Recall Lisa whose early abuse and resultant fear of rejection by an adult had such a deleterious effect on her academic motivation. Transferred into Bill and Susan's classroom with its emphasis on student self-directed learning, Lisa felt "safe" for the first time in her school life. She spent the first few months of school avoiding all contact with the teachers. In her folder, she would describe her learning plans in the vaguest possible terms, and at the end of the day she would simply place a "check" next to her initial plan.

Then one day, she discovered Ali in the computer corner. Ali showed Lisa how to type on the computer and the two girls had a brief computer conversation about a TV show they'd both seen last night. Soon Lisa was spending a lot of time in the computer corner, either in computer conversations with Ali on a variety of academic and non-academic subjects or working with one of the many programs Ali recommended to her.

One day, as Bill was walking past the computer corner, Lisa shrieked with delight, whirled around in her seat, spotted Bill and said, "Look at the Logo drawing I just made!" Bill was so startled at this spontaneous invitation to him that he scarcely noticed the sophistication in Lisa's design. But his smile must have been all the reaction Lisa needed. Increasingly, after that day, Lisa began sharing her work and her questions with Bill and Susan.

The computer had opened the door to communication for Lisa. Yet it was not so much the computer as Bill and Susan's patient educational philosophy and rich educational evironment that made Lisa's discovery of the computer possible. Once the communication channels had been opened, Lisa's motivation to learn accelerated at an incredible rate. Here, although lack of motivation had been a presenting symptom, it had not been the point of intervention; rather communication had been.

CONCLUSIONS

In this book, you will meet many more students whose learning has been facilitated by using computers. Unfortunately, these fictionalized cases represent the very small number of people with special needs actually benefitting from computer use today. Despite the vast potential of computers, social priorities and high development costs have thus far denied these possibilities to most people with special needs. This is not a technological problem. It is a matter of social and political will.

By the same token, not all special needs are appropriately addressed by computer use, nor are computers by themselves ever sufficient to meet educational needs. Thoughtful educators need to work with students, helping to define their needs and determining what adaptations to their environment can most appropriately assist them in attaining these. If the adaptation involves computers, the next task becomes to find the means to supply the needed equipment and materials. If the adaptation does not involve computers but works well anyway, that is certainly an appropriate choice to make. Computers are great assistants; they are not and will never be a panacea for special educational needs.

Communication and Language Arts 3

Weakness in some basic aspect of communication is almost universal among the students served by special education programs. The nature of our measuring instruments leads us to describe these deficits in communication in terms of particular and specific language skills. However, it is the disruption of communication, and not problems with individual component skills, that is of greatest importance.

For pedagogy, the essential feature that differentiates communication from its component skills is its purpose: communicative intent. To teach communication, it is important to respect that purpose. Skills drilled in the absence of autonomous communicative intent may be essential components, but are not, themselves, communication. Does this matter? If people learn the component skills, will they not be able to coordinate them and use them to communicate? For individuals who are filling in gaps or extending their otherwise fully functioning communication, the answer is, generally, *yes*. But this approach does not always make sense for students with serious communication disruptions. If only we could first make them competent communicators and then teach the skills! That task, which seems an impossible paradox, is precisely what the computer may help to accomplish.

The computer can be programmed to recognize even a very limited repertoire of acts and to respond to those acts in predictable and interesting ways. Thus, whatever skills a child has can be used to control the computer, allowing the child to communicate intention to the machine right at the outset. As more skills are developed, they can be used to enhance the communication that is already in process. More options open up, but communicative intent is well established first—the child is being a competent communicator.

Even small limitations in communication reduce one's ability to interact with the world, cut one's choices in activities, and narrow one's experience. Nevertheless, communication is not an end in itself—having an estimable and enjoyable life is. Communication is a means to the life a person wants to lead, and different communication styles are needed for different people, activities, and times. For an academic, reading and writing are essential, but learning to communicate with mime may not be worth the effort. By contrast, an actor must be able to work without words.

Though vitally important, communication's role everywhere—especially in the eyes of a student—is secondary to the activity and autonomy it serves. Unless a communication weakness interferes with some other desire or goal that the individual has, it is of little concern by itself. This, too, must be respected in planning strategies for improving communication.

It is important to note that impaired communication, in addition to impeding the development of some abilities, can mask others that have successfully developed. As a result, children with communication impairments of any sort are often vastly underestimated. Severe communication handicaps frequently give a false appearance of mental retardation, contribute to social isolation, cause severe behavior disorders, and are educationally devastating. The implication for teaching is that there are times when rich, stimulating, communicative activities that appear too sophisticated for the special student are really quite appropriate, and should not be overlooked.

Teachers and educational planners have long valued environments that are rich enough to help students be autonomous and active, and in which they communicate naturally about their activities. Many elementary school classrooms and learning centers successfully use this model. But the weaker a child's communicative ability, the more difficult it becomes to provide the same level of autonomy and control, at least with conventional educational methods. Again, the computer can contribute by being responsive even to limited skills. It immediately expands the range of accessible and interesting activities, and that, in turn provides a student with more to think and communicate about.

Secondary school environments are generally designed differently and tend not to be described in terms of autonomous activity and spontaneous natural communication. Still, the computer's value in the development of language and communication is often much the same as it is with younger children. Departmentalization and heavy class loads reduce the opportunity for secondary school teachers to know students fully. It is unrealistic to expect all subject area teachers to have the time, skill, and sensitivity to integrate special communication needs into their programs. This puts nearly all the burden on the communication specialists. With conventional methods, the frequent result is that communication skills are taught in a vacuum, away from the art, games, science, or socializing that may be the student's primary interest. This approach to teaching communication is second-best. To learn communication by doing it requires communicative intent, a goal of immediate interaction, autonomy, or control. This is where the computer's greatest strength lies.

One Example

Michael is a bright ninth-grader who has a strong interest in electronics, but great difficulty with work that requires much reading or writing. His reading is slow, labored, and unreliable. His spelling tends to be a phonetic rendering of his pronunciation.

One aspect of Michael's educational plan might focus on his interest in electronics. The school offers an electronics course, and has a shop in which Michael could work. The direct involvement with electronics would be great for Michael, but more is needed. Michael would, no doubt, learn to recognize a host of related vocabulary items, but there is little in the course that would help him to improve his handling of a block of text—certainly not enough to help him make significant strides toward learning to read an electronics book. While his reading is so weak, he remains heavily dependent on others for information and guidance, and certain opportunities are closed to him.

Although part of Michael's educational plan might focus directly on improving his reading skills, this directness has drawbacks. Reading is already an area of sensitivity and pain for Michael. He knows he has a reading problem, and is certainly accustomed to the lessons, but it is nonetheless hard for him to sense any progress in those lessons. From one lesson to the next, he can't make practical use of the change that occurs.

What concerns his teachers most, namely Michael's reading, is not what concerns Michael most. Michael has developed sophisticated strategies for coping without reading. His interest is electronics.

The ideal plan would incorporate Michael's language lessons into activities that are meaningful to him without bogging down those activities or trivializing them.

SOPHIE (not yet commercially available) is a novel computer activity that was developed specifically for teaching high-school students about electronic troubleshooting that would be far too time-consuming and hazardous for a student in a real laboratory. Besides being a wonderful electronics teacher, SOPHIE presents a model for combining reading with other studies. The program simulates an electronics laboratory in which the student is brought a faulty piece of equipment to repair. The simulated device to troubleshoot is a regulated power supply simple enough to be studied and understood by a beginner, but complex enough to present opportunity for considerable exploration and growth. Also available in this imaginary laboratory are some measuring devices, spare components, circuit diagrams, and properly functioning models of the same power supply.

The student may ask the computer for an easy, medium, or hard fault to find. The student is then free to "roam around" in this simulation, replacing parts, performing experiments, making measurements on the faulty device or on unfaulted models, or asking the computer to perform computations. Each test reveals a piece of information that may or may not be a clue to the nature of the fault. The student may also ask the computer to offer advice. Such advice might come in the form of commentary on the strategy the student uses in troubleshooting. For example, when the student askes SOPHIE to make a certain measurement the program might first reply that the result can be predicted from tests that the student had made earlier and then help the student to deduce the outcome.

All of the experiments in this simulated laboratory are performed by typing questions to the computer and reading the computer's replies. SOPHIE allows the student to phrase questions flexibly, and programs of this sort can answer back using variety in their English. The program is also fairly tolerant of misspellings. For example, the student may ask,

> WHAT IS THE VOLTIGE ACCROS RESITOR R4

or

> HOW MUCH VOLTIGE IS ACCROSS IT

or

> WHAT IS THE VOLTAGE DROP BEETWEEN POINT 1 AND PIONT 2?

In some cases, *SOPHIE* may rephrase the student's question to confirm its understanding.

Due to the simplicity of the language and the familiar content, Michael can quickly become a competent communicator. Lessons to teach specific language skills can be designed either in the form of modifications to the computer program or as supplements to it. Imaginable modifications include making the program less tolerant of Michael's spelling errors or allowing the machine to use more complex language in the messages it gives to him.

Even without specific tailoring of the vocabulary and text structure to suit Michael's particular reading needs, this activity may be of tremendous value to him, if only by placing reading in a friendlier context. Further, the quantity of reading and writing Michael does in a given period and the quality of attention he pays are both likely to be greater in the *SOPHIE* environment than when he is working with a static worksheet or text.

Temporary and Permanent Communications Aids

Though there is some overlap, aids to language and communication might broadly be classified into those which serve a temporary educational, remedial, or therapeutic purpose (somewhat like training wheels) and those which are permanent supports (like eye glasses). The need for a permanent support is likely to be determined largely by an individual's disability, and the nature of that support largely by his or her ability.

Among temporary educational aids, however, the choice of activity depends more on interests, needs, and history than on ability and disability, and is often quite complex. For Michael, interest was a key factor in the decision to try *SOPHIE*. Different interests or background would have required a different activity, and perhaps even a different approach, despite a similar language problem. On the other hand, a student with an entirely different language problem—for example, limited experience with the English language—might also use *SOPHIE* to good advantage.

CHOOSING APPROPRIATE COMPUTER INTERVENTIONS—LOOKING AT THE DISABILITY

Disorders That Affect Strength, Coordination, or Range of Movement

Cerebral palsy, muscular dystrophy, ALS (Lou Gehrig's disease), arthrogryposis, high spinal cord injury, stroke, arthritis, and other conditions affecting movement can severely limit writing and/or speech. Individuals with these impairments may benefit from a communication aid designed to accompany them throughout their lives.

Often, the educational burdens of these individuals extend beyond the superficial effects of the communication disability. Non-speaking children, especially those who have never spoken, tend to be talked to less often and less intelligently than their speaking agemates, even though they may be fully as capable of hearing and understanding. If they also have other motoric disabilities hampering their mobility or their ability to manipulate objects, they are further isolated from social interaction and from the pleasure and learning that accompanies it. For these individuals, not only is a permanent communication aid required, but educational support to remediate lost environmental learning is required as well.

Both permanent aids and temporary educational aids must accommodate the personal growth they attempt to foster. When a lesson is learned, it becomes a waste of the individual's time if it is continued. When a communication aid is outgrown, it becomes a limiting factor and needs replacement. This is not an idle consideration. The first communication aid that is given to a child is deliberately kept simple to avoid distraction, confusion, and frustration. Yet, if that device is successful, it should cause rapid growth in the child's communication skills with the result that its simplicity soon becomes a hindrance rather than a help to the child. If the aid's growth lags too far behind the pace of the child's communication growth, the result is, perhaps, an even deeper frustration.

Eric, whom you met in Chapter 1, uses the computer both as a permanent communication aid and for his education. He is unlikely ever to be able to speak intelligibly, and therefore some version of talking computer will remain an important part of his life. At the same time, there are language lessons he has already outgrown. When he was four, he played a lot with a computer program called *CARIS* (Computer Animated Reading Instruction System). Although the program was designed for all beginning readers, it had a special importance to Eric. Like many other children, Eric had often been read to by his parents, but unlike other children, Eric had little opportunity to respond openly to the stories as they were read. He could rarely point accurately enough to the tiny pictures, and could never help to tell the story by naming the pictures.

Eric may well have learned to recognize a number of words just from sitting on his parents' laps as they read a story to him, but his first chance to be an active reader or storyteller was with the computer. *CARIS* presented him with a list of words like *Eric, cat, dog,* and *house*, and drew an arrow next to each word, one at a time, cycling back to the beginning of the list and repeating the process after all four had been pointed to. Eric could stop the arrow at a particular word just by pressing a switch. Then the computer would speak that word out loud, and show a corresponding picture on the screen. By selecting from a second list, this time of words like *grows, gets smaller, runs, is flying*, Eric could animate his picture. The computer would speak the whole sentence (e.g., The cat grows, or Eric is flying), and show the corresponding animation.

At age seven, Eric has become more familiar with reading and spelling than most of his able-bodied classmates, as he has had to rely on print for so much of his communication. He showed *CARIS* to a group of the kindergarteners in his school, but he doesn't need it himself any more.

Impairments of Hearing

That hearing impairments may affect speech is obvious. Never having heard speech certainly hinders its acquisition, and even a late-acquired

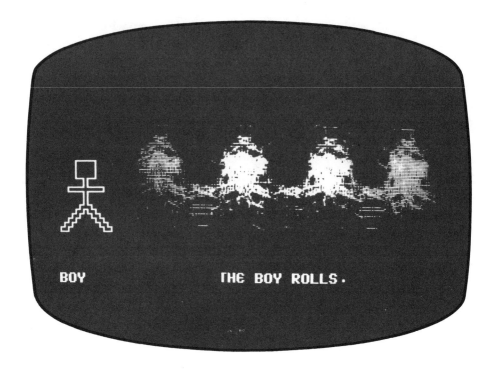

Fig. 3.1 CARIS screen.

deafness reduces one's ability to monitor one's own speech with the common result that the quality and intelligibility of that speech degrades. But a child deprived of hearing from birth loses more than speech. Deafness cuts an individual off from a variety of casual sources of information and language growth, such as the radio and much of TV. It also vastly reduces the opportunity a person has to learn the surrounding spoken language. Thus, even learning to read, which for most children involves primarily the learning of a visual code for a language they already know, can be a significant hurdle for the deaf child who is often simultaneously learning the language for the first time.

One of the major consequences of such a handicap is powerfully illustrated by a problem that Gallaudet College faces. Gallaudet's undergraduate student body, almost exclusively deaf or hard of hearing, is exceptionally bright by college standards. Yet many of these very bright students cannot read a seventh-grade science book because their English is so weak.

These communication problems do not point to the need for a lifelong aid, but rather to an academic educational aid. As *SOPHIE* helped Michael learn electronics, computers can help these students to learn advanced ideas without first requiring them to have mastered the language of the advanced textbooks. At the same time, lessons built into or around these programs can help the students acquire the written language with which to express the ideas they are learning. These lessons also may help to provide enough content and reading skill to allow students eventually to read more sophisticated books on their own.

There are two areas in which computers may play a lifelong role in aiding communication for deaf individuals. Computer mail, which transports written messages instantly through computer networks or over telephone lines, will certainly improve the speed, quality, and ease of written communications for all of us. It may be of particular value to deaf people in serving those aspects of communication for which hearing people depend on telephones. Also, research in computer aided speech recognition, still in its infancy but developing very quickly, has led to experimental devices that can report the words that they "hear" and that encourage dreams of automatic speech-to-text conversion in the future. Even with today's limited technology and knowledge, devices that convert what they hear into visual symbols are capable of enough accuracy to augment speechreading and make it much more reliable. (See under Hardware in the Resources Section.)

Impairments of Vision

Severe visual impairment may, of course, prevent an individual from reading in the conventional way, but it may also affect communication in ways other than the obvious. A blind person may have a variety of problems with writing that are not faced by the sighted individual. Handwriting is hard to acquire when one has not seen a model and cannot monitor one's own performance. But even for the person whose visual impairment began after the acquisition of the skill, or who uses typing to produce text, special techniques are often required to enable that person to review and edit his or her own compositions or even to continue working on it after a break.

The educational issue in this case is, for the most part, the same as the lifelong issue: access. The necessary aids are as valuable outside of formal schooling as they are in.

A large and growing number of computer aids for the visually impaired is now commercially available. (See under Hardware in the Resources Section for details on the following devices.) Talking computer terminals allow blind persons to hear what they type, getting feedback either letter-by-letter or word-by-word, as they choose. The same technology makes it possible to edit a previously typed document. Talking calculators make it possible for that increasingly essential tool to be used by a person who cannot see well enough to use the silent kind. Another machine uses a microprocessor to allow tape recorded speech to be played back at up to 2.5 times the speed at which it was recorded without sacrificing intelligibility. This device gives the blind reader access to spoken text at rates comparable to the reading rate of a sighted person. Finally, there are machines that can read a printed page out loud, whether the page is a telephone bill, a newspaper column, or a story book. These reading machines, especially as they become less expensive, may become an important aid for sighted individuals with specific reading disabilities.

Computer programs have been designed to teach handwriting to blind people by giving auditory feedback on the accuracy of hand movements during the formation of written letters and words. Such auditory feedback on movement also gives rise to a number of audio versions of familiar video games. One company markets such a collection of games including an audio ping-pong!

Access influences what one does and what one learns. Restricted access to books limits one's models of literate writing. That, in combination with the difficulty in editing one's own written compositions, may be devastating to the development of good compositional skill. Devices such as those mentioned above will help remedy the lack of access, but when access has been denied for a long time, special educational interventions may be needed as well.

There are also reading/writing problems not well solved by today's technology. Consider the task of performing arithmetic computations without paper and pencil. Though what is required is often little more than a memory aid allowing us to write down the intermediate results of problems as we work them out, that aid to memory becomes essential as the complexity of the problem increases. A blind student's difficulty with mathematics may stem directly from the lack of a suitably convenient technology with which to keep track of his or her thinking.

This problem affects not only the blind. Ricky, a very bright high school boy with cerebral palsy found himself unable to manage second year algebra primarily because he could not write. A computer program was written for him (by other high school students!) which allowed him to make notes as he worked through a problem, and thereby keep track of his computations. The computer did not perform any of the mathematics; but it became an animated scratch pad on which he could work. The fact that many mathematical problems are expressed with symbols arranged two-dimensionally on the page, however, means that sight remains important. At present, the convenience of Ricky's animated scratch-pad is still unavailable to the blind.

Disturbances of Emotions and Behavior

Emotional/behavioral impairments and communication/language problems are so interrelated that it is sometimes difficult to tell which of the two is cause and which effect. Even if an individual has come to our attention because of emotional or behavioral complaints, it is important to check out the possibility that faulty communication is at the root of the problem.

Children with intermittent hearing problems (e.g., due to chronic ear infections) are often unmanageable. This is not surprising when one thinks about it. Sometimes the child truly does not hear instructions or commands. Other times it is apparent that the child can hear. If the hearing problem remains unrecognized or misunderstood long enough, as it very frequently is, the child may acquire an undeserved reputation for willful disobedience. Once acquired, that reputation has its own way of maintaining itself even after the causal problem is corrected or at least better understood.

Even if the hearing problem had been recognized early, it can become an easy excuse for disobedience at times when the child has, in fact, heard. Further, any diminished receptive ability (due to impaired hearing, neurological damage, retardation, autism, etc.) tends to make the world appear more chaotic and capricious than it otherwise might seem. It causes people to miss many of the explanations of events that have happened or are about to happen, explanations that are available to the rest of us.

Finally, perceived inadequacy in areas that are highly valued by one's parents or group can be emotionally ravaging. Language problems in general—speech delay or defect and reading disability in particular—rank high on the list of failures our society does not tolerate.

Language is also a sensitive indicator of one's emotional state. Some forms of stuttering, extraordinary quietness including total mutism, abnormalities of articulation including "immature" speech, and reading problems are associated with emotional causes. Whether the behavioral problem arises in part from the communication failure (as may be the case of autism) or whether emotional causes lie at the root of the language disturbance, communication and self-expression are often closely entwined with fundamental issues of control. When the handicap being addressed is an emotional/behavioral one, almost regardless of cause, most of the value of the computer will be in educational or therapeutic contexts rather than as a lifelong aid—except, of course, in the general sense in which computers are destined to figure heavily in the lives of all of us. The computer's greatest contribution to education and therapy will probably be its controllability, the fact that it is not chaotic or capricious. Learning to communicate with the computer requires less than learning to communicate with a person; learning to control the computer requires less than learning to control oneself. (More about this role for the computer can be found in Chapter 5.)

CHOOSING APPROPRIATE COMPUTER INTERVENTIONS—LOOKING AT THE STUDENT'S HISTORY

Very Early Impairment

A person who grows up without fluent expressive ability develops a lifestyle adapted to its absence. That person's family and other contacts—though they may not fully accept the communicative disabilities—cope by adjusting their behavior in ways which, at least, make the lack of communication tolerable. Often these adjustments actually reinforce the lack of communication. Thus, the disability may already be a fully integrated part of "normal" life by the time the child attends school. This state of affairs has unavoidable implications for the development of appropriate pedagogy. The intervention issue here is not one of "breaking bad habits" but of recognizing that such habits have, in some sense, not been "bad" at all. The "maladaptive" behavior patterns that we as educators may seek to eliminate are maladaptive only in the sense that they hinder future progress. However, looking at the past, they have served an entirely satisfactory way to create the only life our student knows directly. The key issue, then, is not one of motivating the acquisition of a new skill (or the breaking of an old habit) but of motivating a change in the student's life.

The motivation to give up behaviors that have already served well, especially if that requires great effort, is seldom easy to arouse. The initial task here often involves convincing the person with the long-standing disability that the gain—more varied fluent communication—is worth the effort. Specific skills can be worked on more easily once the utility of the language in their lives is established. (More about the computer's role in this process can be found in Chapter 5 as well.)

An educational plan aimed at establishing the utility of language in the life of the student (and thereby improving communication skills) might begin with music, art, or recreation and may include games that use language, but which do not teach language explicitly. It may also include computer mail which allows one person to compose a message at the computer, address it to a friend, and have the message automatically delivered as soon as that friend uses the computer. Computer message systems tend to be based on text, but could just as well be designed to accept and transmit non-text messages as well, messages composed with **Bliss symbols,** pictures, or even combinations of text, pictures, and voice.

Being Transplanted into a New Culture

Some people are fluently self-expressive, but not in the language of the surrounding larger culture—for example, a person living in Dayton, Ohio, who is fluent only in Spanish or sign language. Their expressive and receptive abilities are fine—they can get complex information put out at an adequate rate, and communication is rapid and easy for them. Their problem is a function of location. The special educational issue is to offer these people freer options, to help them communicate with a wider audience, most particularly those in their immediate environment.

Techniques for the teaching of a second language have become very sophisticated and efficient. Since the advent of the microcomputer, a tremendous amount of software for second-language teaching has also been developed. However, the current limitations of computers must be recognized in selecting among these computer programs and deciding on an educational strategy. Getting a computer to understand and respond to a reasonable vocabulary and unrestricted syntax, even within a fixed topic, is very difficult. Nowhere is any computer capable of handling conversation in which the subject is also free-ranging, as it is in casual day-to-day language use. The conversational ability of a 5-year-old is vastly beyond the best of today's computers.

Some of the methodology of second-language teaching can be handled quite well on the computer, and some of the software based on the latest teaching techniques may contribute to a good educational program. But much of the second-language learning depends on conversation about activities in which one is engaged with others. The strategy of using the computer to provide such an engaging activity may be of considerable help in an otherwise sound educational program. Activities that use language without specifically teaching it, activities that stimulate conversation between human partners, and laboratories designed for experimenting with language may all contribute. A particular advantage in some computer activities is the enjoyment that they can provide using the new language, which may otherwise be associated with failure and a loss of power for the student.

Mary, a Cambodian refugee child adopted by an American family, was learning English while under the extreme stress of having changed her life completely. Her first contact with a computer was a video ping-pong game her family had at home. It served mostly as a source of stability and sameness, making the computer a safe, predictable object. Her school recognized the opportunity and found a variety of computer-based language activities for her including computer-designed word searches, drill programs, spelling lessons, and some games. The drills taught specific lessons, while the games gave her more opportunity to practice using her language. Probably, the greatest benefit of both for Mary was that they allowed her to get her practice with the security of relative privacy. Not being able to communicate as one wishes can be embarrassing and inhibiting. Doing one's initial practicing with a computer removes the worry of being thought slow or silly or being misunderstood by a person.

Impairment Later in Life

Then, there are those whose expressive communication is not rapid, easy, or highly flexible, but who had previously established lifestyles dependent on varied, fluent communication before the onset of their impairment—as in the case of accident or stroke. These people are likely to be quite aware of the advantages of overcoming their new limitation. Although there are often motivational tasks connected with feelings of depression and despair arising from their condition, effort may generally be addressed directly at ameliorating the specific language deficits. Their case is generally less complex than those of people who had never experienced fluent communication. A program for these people must concentrate on restoring the ease and fluency that they once had. Here technology may play the key role in providing the means, but appealing activities can minimize the burden, or at least the awareness of the burden, of regaining lost skills. Since a particular issue with these people is their sudden isolation, computer mail may be especially valuable as a symbol as well as the reality of keeping in touch.

LANGUAGE-LEARNING ENVIRONMENTS

There are hundreds of programs that purport to develop language and communication skills. Many of these programs are not appropriate for use in special education settings. This is largely because the programs often have such limited goals and inflexible styles that they cannot be adapted to the needs of a special student. In smaller part, it is also because most of these programs were not developed with the special student in mind.

On the bright side, many activities that were never explicitly intended to be used for language teaching do contribute greatly to language development. Children programming in Logo, for example, are observed to develop verbally as they describe the behavior and structure of their programs.

These two facts—that much of what is created specifically for language is unsuitable and that almost anything, whether or not it even uses language, could be useful—makes it important to have criteria by which to choose materials and techniques, and a framework within which to use them with special students.

The many ways in which computer-based activities can support language development can be conveniently classified into four types. Three of these are discussed in this chapter: Language-as-Currency (for example, *SOPHIE*), Language-as-Subject-Matter (for example, a spelling lesson), and Tools-for-Communication (for example, word processing and Eric's talking computer). The fourth, consisting of activities which are not, of themselves, directly language-related at all, but which are exciting enough to generate intense involvement and stimulate personal interaction and communication, is discussed in depth in Chapter 5.

Each type of activity creates a particular environment for the student, has its own philosophical and psychological basis, serves certain specific needs, and is best used to promote particular facets of communication. What is appropriate for a student is likely to change as the student changes, and must be reevaluated regularly.

Before examining each of these environments separately and in detail, it is useful to see how all three may interweave in an educational plan.

A Complex Language Learning Environment

Mary, the Cambodian refugee child, was 9-years-old when she joined her adoptive family. The early months were especially difficult for all. Mary was still being treated for medical problems and was having a hard time adjusting emotionally to a new family, culture, language, surroundings, and the immensity of her new city, all at the same time. Her family took the unusual step of studying her language while she was beginning to learn theirs, and this pleased Mary greatly, as she had a chance to do some of the teaching, and to enjoy their readiness to struggle with something that was already easy for her.

The family owned a home computer which they bought partly for business purposes and partly for their older daughter Kimanne who had just entered high school. Kimanne used the machine for typing and editing her papers, but also had a real flair for mathematics and became a good programmer just to play more with the mathematics. Kimanne actively disliked the few games that the family had bought for the machine, but showed Mary one ping-pong-like game with which she instantly fell in love. It was a one-person game requiring no reading, typing, listening, or talking, and Mary played it for hours, alone. For a long time, she showed no interest in any other games, or in having anyone play that game with her, but eventually, she invited other members of the family to join her. She always beat them! The computer's role here was primarily one of emotional support, but it did serve communication development directly in one way. It was an activity around which the social chatter was focused, and unlike any other activities, this one was Mary's turf. She was the central figure.

Mary began trying the other games, but stayed mostly with the first one. For a birthday, Kimanne gave her a new game. The player searches through a maze of caves looking for an imaginary beast and avoids being carried away by bats or falling into pits. It was a very different sort of activity for Mary, and a risky one, but it belonged to her and she loved that. Mary liked the pictures it drew, and enjoyed the short, simple repetitive messages that it gave. She didn't play systematically, and she often got eaten by the beast, but she loved to play.

Again, probably the greatest benefit to Mary was the stability and predictability this game represented in her life which had at times been chaotic. But this game added another benefit. Unlike the ping-pong game, it is played entirely in English—simple English at which Mary could become totally competent. This language-as-currency activity added to her practice with English and to her feelings of accomplishment with it.

Her school capitalized on her attraction to computer activities. She resisted situations that had too much variety, unpredictability, or ambiguity in them, and so didn't like the so-called "adventure games" at all. (Adventure games are computerized fantasy exploration games similar to the popular "Dungeons and Dragons." One such game is described on pages 66–70.) She did like games in which she could know in advance what she was looking for and how to look for it. She loved word-search puzzles, and a computer program helped her teacher create reams of these puzzles using words Mary knew. Mary also liked some of the drill and practice activities on the computer.

The school orientation at this stage was language-as-subject-matter; activities that were chosen were intended to practice or refine specific skills that Mary had already acquired through less formal means. Mary was learning bits of English from the computer. Still, it is likely that the greatest value the computer held for her was its simplicity, regularity, and predictability. The computer exercises were easier than free communication and gave Mary a chance to excel easily.

As her English became good enough, her teacher had her try *Textman* and *Storymaker*. *Textman* is a game in which the computer presents a series of story parts, and the object is to arrange them into a coherent story. *Storymaker* also presented story parts to choose among, but here the choice was entirely up to Mary. Mary enjoyed *Textman* but, surprisingly, liked *Storymaker* even more though she had in the past gravitated to activities that had single, predictable, correct answers. From her very first contact with *Storymaker*—the selection of a story title from a list—she had to make decisions about which way the story would turn. At each decision point, *Storymaker* offers several already written options for continuing the story. For each optional twist of the story, the program has an appropriate set of new options for continuing the story. Mary played with each of the stories several times, changing their course, and printing out the story she had thus constructed.

Storymaker also provided a way for Mary to write her own continuation instead of choosing any of the ones that had already been written. Although Mary had been told about that possibility, she didn't try it until she had practically memorized the already existing story parts. In fact, she didn't even seem to remember it was possible until she noticed a change in one of the stories, asked why it was there, and was told that another student had written it. The fact that she could write something that would become really permanent, part of the machine, a story part waiting there for the next child to use, meant a lot to Mary. She wrote neatly tied-up endings for all the stories, and checked each day to make certain her work was still there.

Both *Textman* and *Storymaker* may have enhanced and developed the complex skills of using the interrelationships between sentences, but *Storymaker* had the clear additional value of giving her a safe vehicle for public statements. Mary's writing was an outlet for her feelings, and her feelings motivated the written work that clearly aided her language development. The computer's role here was significant and complex. *Storymaker*, designed very much as a game in which language is both subject-matter and currency, had a strong emotional impact on Mary, and provided tools to support the communicative efforts it stimulated. Thus, for Mary, it combined all four types of language-learning environment.

Following a design that she read in a book, Kimanne wrote a set of Logo programs that allowed Mary to teach the computer to make up silly (but grammatical) sentences with words of her own choosing. This language-as-subject-matter activity differs from the skill drills in that it encourages exploratory learning in which language is the subject of an experiment, much as electronics was the subject of experimentation in *SOPHIE*. Because Mary could choose vocabulary that was loaded and meaningful to her, some of the computer's random sentences were very exciting to her, a common reaction of children to this program. Kimanne taught Mary how to use the word processor to type up a story based on some of the sentences that she had printed out. The word processor became practically a friend to Mary, a book into which she could literally pour her stories and a tool with which she could make them beautiful.

It had been clear for awhile that writing would become an easier emotional outlet for Mary than speech, but the importance of the computer in facilitating that writing, and in leading Mary to it in the first place might not have been predicted. What began as a skill-oriented game that Kimanne wrote for Mary led to an activity and skill that may affect the rest of her life. Certainly, the word processor is a sophisticated language tool that Mary is likely to use far beyond any of the other computer activities in which she has engaged.

THE LANGUAGE-AS-CURRENCY MODEL

The key feature of this teaching model is the focus on pursuits other than language while making extensive use of the student's communication skills directly in the service of those other pursuits. Many computer simulations present their outcomes as text and some—*SOPHIE*, for example—require a fair amount of text as input. Few of these activities were created with language development in mind, and yet, they provide just the sort of language exercise some students need.

Despite how natural it would seem to create simulations on the computer, despite the obvious power of these simulations in education, and despite the existence of clever simulations in biology, genetics, economics, physics, and political science, to name just a few areas, simulations are not yet a widespread commercial item. Still, some are available from educational publishing companies and local computer stores, and more are being produced at levels suitable for elementary school children as well as for older students. See any of the numerous popular computer magazines for descriptions of simulations created for other purposes but which can form a part of a language education plan as well.

SOPHIE's ability to accept misspellings and varied input is still beyond most commercially available packages, but the sophistication of commercial software is increasing rapidly.

Such environments are ideally suited to addressing communication's purpose. There is enough variety among available activities to permit a student with poor communication skills to function as a competent communicator about something that is personally meaningful and important.

Competence as a communicator can be achieved in part because the total universe of language required by such programs is more limited than that required in free conversational speech, and in part because the very structure of the material is interactive. Each block of text is the result of an experiment initiated by the student in a context about which the student already has considerable knowledge. This interactive structure can help the student to understand passages which would remain mysterious had they appeared in a static text. It is best if, in addition to these other advantages, the program's text can be modified and personalized. Judicious modifications can become the basis of language lessons built into the interaction. However, lessons designed around the computer interaction that are not built directly into it can also supplement the education plan. These supplements might range from skill training sessions to free-flowing conversations stimulated by the computer activity.

Another kind of interactive computer environment is inspired in part by the increasingly popular commercial game *Dungeons and Dragons*. This game creates a magical fantasy land in which the player becomes an intrepid adventurer. In some adventure games, the player explores through forests and fields, and down into dark dungeons where treasures, dangers, and other adventurers are to be found. Secret passages lead to other strange worlds. An inflatable raft found in a dark room far beneath the earth becomes essential for a journey along a creek in the mountains up above. These games can be such complex fantasies that even though their structure may remain unchanged, their possibilities seem to grow with the player. Enough people have found these games so intensely exciting and intrinsically motivating, appealing to their sense of adventure and fascination for magic, that variations have proliferated under a variety of names suitable for a wide range of reading levels, and are sold by practically every computer store and many publishers.

For language development purposes, adventure games and simulations are very similar. To decide among the alternatives, the most important consideration is the student's interests. With Michael, either would have worked, and the choice of *SOPHIE* had the added benefit that it gave him another area of special competence, electronics. A general consideration in choosing adventure games is the nature of their politics. Some trade openly in genuine violence. Some are racist and sexist, some subtly and some rabidly. But some are clever, exciting, and relatively benign (slaying, at worst, a dragon).

As with all computer activities, there are practical considerations as well. One ingenious variation on the computer adventure theme is *Snooper Troops*, a light mystery in which groups of players can cooperate as the detectives, trying to catch whoever-done-it. Clever messages, attractive graphics, and the same sense of free exploration await the detectives. However, the designer of the program deliberately chose time pressure as a way to encourage cooperation among students. Events occur and you cannot stop them or slow them down. If you are not a fast reader, you just miss what is being typed to you. Students must work in pairs or small groups to catch all the information before it flies by. This may not be fatal with most students; it may even, as intended, encourage a cooperative rather than competitive game-playing attitude. But for some language-weak special students, it may be unhelpful pressure and unusable.

Activities using language as their currency, like the ones described, may have no educational value to a non-communicating or, perhaps more correctly, an idiosyncratically communicating autistic child. For that matter, they are of little worth for any individual who had not already achieved some prior skill in reading and writing English. If you cannot read, you cannot use, and therefore cannot benefit from, these activities. However, one need have neither an extensive vocabulary nor the syntax of Standard English in order to enjoy and benefit from such activities. Language-impaired children with at least rudimentary English reading skills can be competent communicators in an environment they care about because of its intellectual or recreational value. This changes the communication specialist's job from one of building communication, to filling in gaps in functional communication. The benefits are manifold. Having a functioning communication system gives the student a framework within which to understand skill development. It gives the student a sense of self, success, and power. And it places the specialist in the position of helping the student with an act that is the student's goal. It settles many of the motivational issues. It also teaches one of the hardest aspects of communication to get across in the classroom—how to use it to get what you want.

Michael is doing quite well, except for his reading. Understanding how to get what he wants using language has never been his problem. But for children like Tammy, the CP girl you met in Chapter 1, it may be the hardest part. Tammy had adapted quite well to her lack of verbal communication and to the patronizing way in which people talked to her. She had to learn how to satisfy her needs without text production capabilities, and her experiences, until recently, rarely gave her much to say. Without the felt need to communicate, communication skills seemed arbitrary and useless if not a genuine threat to her. But with a more experientially rich environment, one over which she can have real, effective control and one about which she wants to communicate more, the payoff for gaining new language skills is great enough to overcome the high cost (in Tammy's effort) of even the best communication aids we currently have.

Computer Adventures

This annotated excerpt from a dialogue based on one computer adventure game illustrates some of the game's potential contributions to language teaching. The game opens with this statement:

> You are in an open field west of a big white house with a boarded front door. There is a small mailbox here.
>
> **look in the mailbox.***
> The mailbox is closed.

In all cases, the computer responds with a helpful statement in real English and not computerese. The response often incorporates the vocabulary

---

* Throughout this book, boldface type in computer dialogue indicates user input.

of the command, which helps the student understand the response and acquire the structure of English. Both of these features would benefit a student like Mary, if she liked this kind of game. Michael doesn't need the information about English syntax, but the context helps him read the response. The quantity of reading a child does per turn is always quite small, but is enormous when totalled over the course of a game. Reading is essential for the game but the game is fun and so the child is willing to spend time and energy to puzzle out the hard parts. For a student like Michael who is fluent in English but has difficulty reading, this in itself is the most valuable aspect of the game.

Open it, please.
Opening the mailbox reveals a leaflet.

Take the leaflet.
Taken.

Walk southwest.
You are in a dark forest with tall trees on all sides. Tied to a high branch on the nearest tree is a rope which dangles almost to your waist. On the ground there is a pot of gold.

It would be ideal here to be able to tailor the language to the student, either to aid the poorer reader in starting to play the game, or to enhance skill development by presenting a more advanced student with some more interesting or difficult language. One version might begin: "You are in a dark forest. There are tall trees to your left. There are tall trees to your right." Still another might begin: "You have entered a dark forest and find yourself surrounded on all sides by tall cedars."

Take the gold.
You cannot take the gold. Gold is very heavy and your load is already straining you.

If certain vocabulary items were to be stressed, they could be included in the text. In some cases, they could even become the significant objects, actions, and places of the adventure. There is plenty of opportunity to present information about the items as well. It is also important to consider the flexibility one wants in the input syntax. This particular adventure game would have understood the headline syntax "Take gold," as well as the full version typed above. A more restrictive version might respond to headline syntax by saying "Do you mean 'Take the gold'?" or, more subtly and in another spirit, perhaps "Do you really want me to take the gold?" This style of conversational questioning is a particularly effective way of modeling the correct use of the language while remaining an interactive adventure. The machine is not giving the student an error message. It is trying to carry out the wishes of the student and needs to clarify the intent.

That's mean!
I don't understand the word "mean."

The computer cannot understand everything, but it can respond humanly and in English even then.

Climb up the rope.
You try to climb the rope, but alas, you have too much baggage with you. You will have to leave something behind if you are to be a successful climber.

What do I have?
You are carrying:
   A heavy brick
   A knife
   A leaflet
   A lamp
   A sword
   A glass bottle with some water in it
   A brown sack
   A jewel-encrusted egg
   A skeleton key

Put down the brick.
Brick: dropped.

Demonstration of comprehension of the earlier messages, organization and evaluation of information, formulation of plan, and judgment—all important communication skills—are rewarded here and in the strategy shown in the next few moves.

Climb the rope.
You climb to a high branch and see near you a silver bird's nest.

Put egg in nest.
As soon as the jewel-encrusted egg touches the silver nest, the shell shatters into a billion fragments. A jeweled bird emerges, winks at you, and flies northeast, toward the house. As it flies, it often looks back at you and sings a haunting song.

The player synthesized a lot of information to think of placing the jeweled egg in the silver nest, and received a lot of information as a reward. The loss of the egg may have been a penalty for this experiment, but then remembering the information may be even more valuable than the egg that is now shattered. There is no penalty for not being experimental, except that the game is less interesting.

Climb down.
Dark forest. There is a pot of gold here.

Take gold.
Gold: taken.

A program of this sort could be made sensitive to mechanical details such as capitals and punctuation, but as stated earlier, this kind of environment is suited primarily to dealing with the *purpose* of communication. Too many requirements on its form, at least for the beginning player, may interfere with that purpose.

Walk northeast.
Open field.

Look around.
You are in an open field west of a big white house, with a boarded front door. There is a small mailbox here.

Walk north.
You are facing the north side of a white house. In one corner of the house there is a small window which is slightly ajar.

Open the window.
With great effort, you open the window far enough to allow entry.

Climb in.
You are in the kitchen of a white house. A table seems to have been used recently for the preparation of food. A passage leads to the west and a dark staircase can be seen leading

upward. To the east is a small window which is open. On the table is a battery-powered lantern.

Walk up stairs.
It is pitch black. You are likely to be eaten by a grue. You hear a haunting song.

The player may remember having seen a lantern on the kitchen table that would be of use and may go back down to get it. Using language in a purposeful way, controlling with it rather than being controlled or limited by it, is a very powerful experience. The player probably does not yet know what to do about hearing a haunting song, but may recall hearing it just after freeing the jeweled bird from its egg, and may surmise that there is some importance to hearing the song again.

This game and others like it have grown up spontaneously for the amusement of computer buffs. On the whole, their form has not been influenced by pedagogical or other educational considerations. Most are very restricted in the language they can accept as input. Some, for example, require that all inputs be typed in headline syntax like "Take gold." The game in this example understands full English structures like "Take the gold" but is equally tolerant of headline syntax. Commercial games are essentially unmodifiable unless they have been deliberately designed to accommodate that kind of change. Accommodating modifiable versions is certainly within the capabilities of the popular personal computers. And, even if no one working with a child can tailor a game to the child's syntax or vocabulary or the exact reading or writing skills the child's teacher is concerned with promoting, the game can still provide an environment within which the student is competently communicating. If the game is selected so that a child can relatively easily understand essential instructions and responses, is intrigued, and is practicing some needed skills such as reading and writing using consistent syntax, these activities can provide some students with a good interactive language experience. If the activity is rich enough, then the better the child is at understanding the subtleties of the communication, the more interesting the game is.

Precisely because computers are not great conversationalists and do not understand everything that is typed to them, players recognize easily that they must occasionally rephrase their demands and experiment with the language. Children often become fascinated with the very fact that there are words that they know which the machine does not.

Criteria for Choosing Language-as-Currency Activities

In order for such an activity to fit in the environment at all, it must stress communicative function rather than focus on form. In order for it to work, the student must recognize the key words or be able to learn them quickly. It must be controllable (for example, not requiring faster reading or responses than the students are capable of). If the game is not modifiable, you may sometimes want to choose one that operates at, or just slightly beyond the frontier of the child's current ability, with vocabulary and syntax matching the child's next step in development as closely as possible. To make it cost-effective, it must be usable by a number of children, or over a period of time by a single child. Finally, if the child is to stay with it, it must be fun to play.

THE LANGUAGE-AS-SUBJECT-MATTER MODEL

A large proportion of school software follows a drill-and-practice model of education. There are spelling drills, vocabulary builders, punctuation exercises, and all sorts of electronic workbooks. For some children, there is comfort in predictable tasks in which errors are minimized by the measured pace and repetitive nature of the task. That was true for Mary. Drill and practice on a computer is not essentially different from drill and practice in any other form. The computer can manage it closely, can present snazzy graphics as "rewards," and can keep records of the student's progress, but there is little in the electronic workbook or the $2000 flashcard that is a fundamental improvement over the paper kind.

The silly-sentence-builder that Kimanne wrote for her sister represents another kind of "errorless learning," as does *CARIS*, the animation program that Eric used for awhile when he was first learning to read. These programs encourage exploration, unlike drill-and-practice activities, which tend to reward repetition of a restricted set of behaviors. Because the activities are exploratory, they do not need to evaluate performance in terms of success or failure, right or wrong. There are no errors to make. Even so, for a child for whom language has been associated with failure in the past, it is often better to begin with tasks in which language is not the focus, but merely the vehicle of the activity—language as currency rather than language as subject matter. As the child's confidence as a communicator grows, it may become more suitable to move to or, at least, alternate with activities in which attention is drawn directly to language.

In many ways, exploratory language-learning activities are laboratories for experimenting with language, much like *SOPHIE*, except that their subject matter is language, itself. Unlike drill programs, they provide opportunities for learning that are very difficult, if not impossible, to provide without the computer.

Exploratory language-as-subject-matter activities depend on a fair amount of knowledge of the language to be explored, regardless of the child's ability to communicate in other ways. They are of no value for the child who is just beginning to learn the language, and are most suitable for children who already have reasonable skill, but for whom the goal is to enhance that skill, or to raise their awareness of words, spellings, or structures. The following illustrate the range of these activities.

Sentence Construction

Cheryl, a 12-year-old poor reader with serious academic problems, used a sentence-builder program identical to Mary's to produce a gossip column about the children in her class. She "taught" the computer two lists of words, one a list of the first names of all of her classmates, and the other a list of actions that she thought would be useful for a gossip column (e.g., kisses, sits on, hates, makes out with, bites, loves, hits). She then "taught" the computer her definition of a legal sentence: an element of the name list, an element of the action list and another element of the name list. By instructing the computer to compose and print these sentences repeatedly, she generated long lists of gossip (sentences like "Jason kisses Marc," "Susan loves Andy," and "Sandy makes out with Sandy"), literally enough to paper her walls!

Cheryl had been kept back a year, and was still academically at the bottom of the heap in her city classroom, several years behind in reading ability and in written composition. What is striking about her play with the sentence generating program was that she carefully read through all of the sentences that were produced, starring the sentences that she liked ("Marc loves Cheryl") and crossing out the ones "it got wrong." She was able to approach a traditionally academic activity (sentence construction), albeit with a very limited, self-selected, and repetitive vocabulary, without being and feeling like a failure. She was able to apply knowledge of the function of noun and verb phrases in sentences—knowledge which she clearly had but could never demonstrate in class—to teaching a computer how to talk. Although she clearly understood the randomness of its sentences and understood her own role in instructing it in its behavior, she was also able to take the psychologically valuable opportunity to "correct" the machine when it made mistakes. This gave her the chance to take an active, positive role, and may have been even more important than the specific academic skill she learned. She was also able to acknowledge her own investment in the content of its productions even though she was otherwise somewhat shy about saying, for example, that Marc was her boyfriend. Her experience, overall, was much more one of control and communication than of mere exercise or academic effort.

When children discover, as did Cheryl, that some lists of words do not "go together," they may begin manipulating syntactic or semantic attributes of these words to adjust their sentences. For example, Cheryl's list of verbs originally contained entries like "love" which is inappropriate for her third-person-singular subjects, and "reads" which is inappropriate for the choice of sentence construction because she always supplied a person's name as an object. The next step for her in her gossip column might have been to build in some checking to allow greater flexibility in the productions. She chose instead to change all of her verbs to the third-person-singular form. Had it been desirable, the teacher could have helped her figure out how to make the machine conjugate the verb properly depending on the subject. In that more general case, Cheryl would have had to leave all of the verbs without a final "s," and look at the subject for clues. If the subject were I, you, we, or they, she would leave the verb as is. Otherwise she would append an "s" to it. Note that, in content, this is very much the same as a drill-and-practice lesson on the same material, but the spirit and psychological characteristics are vastly different. When errors occur, the child is the only one who can correct them, and it is the "machine's fault, (dumb computer!)." For the normal child, much of the underlying knowledge is already in use by the child, and the major learning involves becoming aware of what is already known and making that self-consciousness available for enhancing the knowledge. For the language-impaired child, this may not always be the case, but at least the communicative content of the exercise will be the product of the child's interests and needs, and thus will have salience to him or her.

Mike Sharples, at the University of Edinburgh, Scotland, developed several computer-based exploratory language learning activities in Logo. Some of his programs are similar to the one described above in that they are capable of replacing structural elements of a sentence (e.g., noun phrases or verbal constituents) by instances of that element (e.g., "loves"), while others are able to recognize words or patterns of words in a sentence (either literally or by some specified attribute) and manipulate them according to a child's instructions. These are not now commercially available, but a teacher who knows Logo very well can write similar activities. One of Sharples' programs can be taught a list of words and be told to delete them from any sentence the user gives to the computer. Using one of these programs, Sharples developed a game which he reports his students particularly enjoy. In the game, he gave the computer a moderate-sized list of common adjectives and adverbs. When a student types a sentence such as:

The little kitten crawled quietly under the green fence.

the machine looks up each word of the sentence in its lists, and then replaces all adjectives it finds in the sentence with an asterisk and all adverbs with an exclamation point, thus producing:

The * kitten crawled ! under the * fence.

The children love competing "against the machine," trying to sneak in rich and descriptive adjectives and adverbs that the machine does not know. Depending on the vocabulary taught to the machine, that can be quite a challenge for many students. In addition to the vocabulary challenge, children are practicing creating more colorful constructions than the degraded form of the sentence "The kitten crawled under the fence."

Story Writing and Reading

Storymaker, the program that Mary loved, has two modes in which it can be used. Besides choosing freely how the story should proceed, the child can choose to be given a goal—a direction in which to push the story. In that case, each set of options becomes a kind of puzzle requiring one to think what continuation would be most helpful in attaining the goal.

At the basic skills level, this kind of play with language is great practice in reading with comprehension. As in the adventure games, the student is faced with very little text at a time. The activity is not a test. Just as the child may add new elements to a story at any decision point from the title on, so may the teacher. Thus the vocabulary, syntax, story content, and structure can be completely and easily personalized to the child's interests and language abilities. These personalizations can have impact at a level deeper than linguistic exercise. Constructed thoughtfully, an imaginatively branched biography of the child can provide opportunity for the child to play with alternative histories and futures in the context of playful fantasy. Moreover, the power to determine how a story ends is a wonderful experience of control, and probably one of the stronger motivating forces behind children's fascination with this activity.

It is worth calling attention to the difference between the philosophy behind the common implementations of branched-plot stories, and the philosophy proposed here. In many commercial branched-plot books and computer adventures, the reader makes a choice of immediate action and suffers some unanticipated consequence from it. As in much humor, it is the aborted threat (the real you doesn't get eaten by a real dragon) that tickles. But even threats that are aborted betray the fundamentally competitive model of these activities—you against the machine, one false move and you're dead—which, for some individuals in some situations, is not desirable. The player is offered control, but the risks are high, and you have no alternatives other than those provided by the machine. *Storymaker* preserves the humor that comes from reading the alternatives written by others, but contains no threat at all. Its ability to accept student-written continuations at all choice-points means that one is never forced to accept unwanted consequences.

By manipulating the story's elements, children learn about the structure of a written composition. Although there is no constraint on the story structure when they write their own continuations, they are, even then, encouraged to consider the structure at least in deciding whether their contribution is intended to end the story or can be continued in a next step.

Even more flexible than *Storymaker*, although otherwise similar in philosophy and pedagogy, is the *Interactive Text Interpreter* (*ITI*), a tool that enables a teacher or student to generate branched-plot interactive texts to suit any purpose. A half-dozen or more interactive texts are available that prompt and develop writing in narrative, expository, poetic, and other styles, and others can be created easily by student or teacher. A pedagogical principle reflected in the prepared texts is that the responsibility for a finished text shifts gradually from the computer to the student. At first, the student may add only a few words to personalize an otherwise fully completed text; later the student makes more substantial additions of a phrase or sentence at a time, but still to a pre-existing story structure; eventually, it is the student who writes the bulk of the narrative surrounding an outline supplied by the machine. By using InterLearn's *Writer's Assistant*, students can then edit their compositions to make further additions and alterations. The *Writer's Assistant*, a word processing system designed for elementary school students, provides a command that allows a student to rearrange the text to begin each sentence on a separate line, thus visually highlighting certain aspects of sentence structure, such as punctuation, capitalization, and overall length.

By manipulating the structural elements of a story or inserting sentences into otherwise coherent text, children are playing with grammatical relationships that cross sentence boundaries. Some of these intersentential grammatical relationships are even more important to skills such as reading comprehension than are many intrasentential grammatical relationships, such as correct formation and use of plurals or subject-verb agreement. Curricular planning that does not recognize this may be inefficient or counterproductive. The goal, for example, of helping a deaf student to become a fluent reader may be seriously hindered by channeling too much of his or her time into aspects of grammar that are observedly weak but not essential to comprehension. The image to keep in mind is the brilliant foreign scientist or statesperson who speaks English with a strong accent and consistently mangles a few grammatical rules, but who can, in English, clearly read literate and sophisticated written materials and can deliver an articulate and forceful speech.

Another activity which deals with whole texts in *Suspect Sentences*, a computer-based game that encourages the players to think about aspects of style in writing. A player chooses the role of forger or detective, and then selects from a menu of writing styles. As forger, the player is presented with a fully written paragraph. It is then the player's option to compose a new sentence and sneak it into the paragraph anywhere, with the goal of making the forged sentence look so much like it belongs that another player who chooses to play detective will not recognize the forgery. Thus the forger is trying to add a sentence to the paragraph without disrupting the style, coherence, or sense of the paragraph. The detective is trying to be a good editor, to recognize which sentence could be omitted from the paragraph without sacrifice.

Speaking and Spelling

Anna is a 12-year-old whose file describes her as an educable mentally retarded child with a variety of perceptual and linguistic deficits. Though she appears motorically intact in most situations, close observation reveals a slight hand tremor and some other subtle signs of neurological damage. Her speech is severely impaired. She rarely says more than two or three words at a time and the sound of her speech is very distorted.

She enjoyed taking a turn as "speech therapist" to a computer that was capable of learning a sequence of phonemes and stringing them together into "impaired" but recognizable speech. At times she was pleased and amused enough just to respond "No!" when the machine asked if it had pronounced something correctly, but at other times, she was eager and able to give specific criticisms, and showed that it mattered to her that she could display her competence.

The machine had a list of words it could already pronounce well, including *hello* and *I*. Anna wanted to teach it to say hello to her teacher, Barbara, and therefore had to instruct it on the pronunciation of that name. She had already learned how to make the computer say the sound *ar* as in *c ar*. Anna recognized this sound in the first syllable of Barbara and taught the machine to say that syllable *b ar* correctly. She had also already learned from experimentation that the machine often pronounced things wrong even if they were spelled "correctly." For example, it always pronounced the letter *a* the way it is pronounced in the word *cat* and so, to get it to pronounce her own name correctly, Anna had taught it that

"Anna is pronounced *a n uh.*" Still, she was often very strongly influenced by the spellings of words and told the machine that "Barbara is pronounced *b ar b a r a.*" Of course, the last two *a*'s sounded like the *a* in *c a t*.

With a gradually broadening smile, she remained silent for a moment, wrinkled her nose, and then imitated the machine's *a,* commenting in her taciturn way that it had said *a* when it should have known better! Printed on the computer screen that she was facing was the question, "Did I say that right?" to which she spoke and then typed "No." When it said "Teach me again," she typed *b ar b a r uh,* correcting one of "the computer's" two errors.

In precisely the area of her most obvious deficiency she was able to demonstrate true proficiency. She assumed a teacher rather than student role, with apparently both cognitive and affective benefits. Spelling might become an experimental-constructive activity in much the same way. A group of fourth- through sixth-graders with whom one of the authors of this book (Paul Goldenberg) was working in a spelling clinic had to find some way that they could learn to distinguish between *ow* and *ou* spellings of the vowel they heard in the word *sound.*

The group began by thinking up a bunch of words with the *ou*-sound and writing them in rhyming lists, using Dr. Goldenberg to check their spelling. One miscellaneous list began to collect the several *ou* words that had very few rhyming partners. Only once did the group discover words that rhymed with each other but contained different spellings for the vowel sound: *flour* and *sour,* which rhymed with *flower* and others like it. The students listed these according to their spelling and not to their rhyming group.

COW	BROWN	GROWL	BOWEL	PROUD	ASTOUND	POUNCE
CHOW	CLOWN	SCOWL	TOWEL	CLOUD	MOUND	OUNCE
BROW	TOWN	PROWL	VOWEL	LOUD	ROUND	MOUTH
NOW	FROWN	OWL	TROWEL	SHROUD	FOUND	COUCH
HOW	CROWN		POWER		AROUND	OUCH
ALLOW	DROWN		FLOWER			HOUSE
MEOW			TOWER		etc.	SOUR
						FLOUR

When they began to study the lists that they had made, they noticed that the vowels in any one of their lists were all, without exception, spelled the same way and therefore began to look for the differences between the lists.

The first rule that the group proposed was that the vowel sound was spelled *ow* if it came at the end of a word, or was immediately followed by an *n*, *l*, or *e*. Later, when they looked at words like *round* and *ounce*, they revised the rule a bit, saying that the *n* had to be "final." If the vowel sound was not at the end of the word, and was not followed by an *l*, an *e*, or a final *n*, it would always be spelled *ou*.

Eventually they found four exceptions (*noun, crowd, foul,* and *powder*), but by then, they had already discovered an amazingly powerful generalization, a trivial two-step rule that could be taught to a computer and that correctly spells 94 percent of words containing the *ow* sound! It was also easy for them to teach the computer to recognize the few exceptions and make it a perfect speller.

All the *thinking* about the problem was done entirely away from the computer. The significance of the machine for this activity is that it provided the students with an environment in which to use the systematic knowledge that they were gaining, a place to try out their model of the spelling rule to see quickly what it produced, and a ready student to teach.

Criteria for Choosing Language-as-Subject-Matter Activities

Language may be the subject matter of formal lessons, drills, competitive games, and exploratory activities, each with its potential applications. That last category, at present the least widely known of the four, is differentiated from the others in that it allows language to be the object of exploratory free play, much as it normally is for the young first-language learner. Some exploratory learning activities may share characteristics with games in which performance is evaluated in points won or lost in competition with another person or with the machine, or even with drills and activities which have a fixed-task/fixed-response or time-pressured structure—but their defining characteristic is that they be essentially exploratory in nature.

Because the computer's role in such exploratory activities is to provide the child a relatively easy and powerful vehicle for applying knowledge he or she already has, this kind of environment is ideally suited to the purpose of consolidating existing knowledge and extending this knowledge into new territories. Like other exploratory learning environments, it lets the child be the experimenter, the teacher, the one in control; it allows the child, within limits, to determine the course of the activity.

When is drill appropriate? When the *function* of communication is well established, and isolated forms are in need of improvement, drill may be beneficial. However, when communication is not functioning adequately or when the child is still learning its purpose, it is important to remember that, although drill might help train some fundamental skills needed for a particular mode of communication, not all uses of these skills are communicative. Thus, the appropriateness of a drill program depends very much on the needs and goals of the child and the child's teacher. Tutoring and repetitive drill outside of an experientially rich and child-controlled environment may conflict and compete with the function of communication. A child for whom very basic concepts of communication are lacking—shared assignment of a particular meaning to a particular symbol, means-end relationship such as affecting another person's behavior by means of one's own—cannot use such drill systems and might benefit more from a Logo-based (or other) responsive environment. In order to develop communication, itself, or even just the motivation to communicate, drilling is not appropriate. Child-initiated and controlled activities are essential as a large part of the language learning experience.

Linguistic flexibility makes exploratory language activities particularly effective as supplements to other classroom experiences. In general, commercial adventure games and simulations are not designed to make it easy to tailor their content, vocabulary, or syntax. One selects what seems most appropriate from among the available publications and then makes the best of it. On the other hand, programs such as *Suspect Sentences*, *Storymaker*, and *ITI* are designed to be personalized. Thus, the pieces of a thoughtfully branched account of the troubleshooting of a piece of electronic equipment might constitute an interesting structure for Michael to play with and expand in *ITI* or *Storymaker*. By creating other branches and their consequences, Michael is building up an instruction manual for the repair of the equipment. The model for this is already present in the story that he first encounters, and the new content comes from his continued explorations with *SOPHIE*. Neither Michael nor the teacher can modify *SOPHIE*'s language much, so it has neither the richness nor variety of running prose, but Michael encounters and even develops that by writing a branched book in *Storymaker*. Similarly, he might try sneaking forgeries into paragraphs about electronics in *Suspect Sentences*, though, in this case, much of the fun would require that another child be as involved in the electronics as in the language. Of course, exploratory language environments can be adapted beneficially from all kinds of simulations, adventure games, or other language-as-currency activities.

TOOLS FOR COMMUNICATION

Computer tools in a language-learning environment can help the child accomplish self-selected goals. By providing appropriate tools for manipulating language, language-learning becomes a by-product of accomplishing a desired task.

Word Processing

Another way in which the computer may serve to exercise and improve already existing language skills is exemplified in the writing laboratory concept. A computer-based writing laboratory ideally provides a number of tools for the manipulation of written language. A well designed word processing system can be an invaluable tool for the young creative writer, just as it is becoming indispensable to the adult business and professional writer. Most of this book was composed and edited at a computer, from the very first draft. Because of the ease of making changes, this chapter, for example, has passed through over 200 "drafts." With paper-and-pencil technology, there is a serious penalty for noticing ways to improve a piece of writing: making the improvement defaces the current draft and forces one to rewrite (or retype) the paper; yet not making the improvement means settling for less of the quality that one can achieve. The word processor enables one to alter a piece of text by typing merely the alteration. Unaltered text need not be retyped. The computer takes care of all the other adjustments and prints out an unblemished work immediately. Thus, there is no penalty for improving the draft of a composition.

Word processing is, of course, wonderful for the active language user who is ready to produce, for the person who has skills but needs a tool to let them flower. But it is also of great importance to the person who is likely to be having special difficulty constructing a good composition, because it allows the revisions to be done easily and without penalty.

Thus, an individual struggling with English as a second language, a deaf student, or a student with a severe spelling difficulty may all benefit from the ease of composing their ideas first, and then fixing their expression of those ideas later. There is much more to be said about word processing and the teaching of writing, but that is the subject for another book in this series.

Computer Mail

Computer mail is also an unconstrained interactive written language activity on a computer. Members of the computer-using community can send private messages to one another and keep files of the mail they send and receive. At one school for the deaf, the children were taught how to send and receive mail. Some teachers wrote mail to the children in the hopes of providing interesting reading material and stimulating written response, but for many children no special stimulation was needed. Great volumes of mail, much of it in the form of love letters and insults, were sent back and forth among the children. Some of the most active mail users showed very considerable improvements in their English-writing ability over the course of the first year using the mail program. (An account of this development, and examples of the letters sent by the children can be found in *Special Technology for Special Children*.) In a similar experiment by Jim Levin and his colleagues, a computer mail link between children in California and children in Alaska served as a focal point for a series of computer-based writing activities. Designed for microcomputers, their mail program and the school-to-school linking capabilities are now commercially available as *The Computer Chronicles Newswire*. This interactive human communication has great social and academic benefits for any child who is capable of using written language at all.

Communication Aids

Taking seriously both the goals a person currently has and our concern for expanding that person's horizons requires us to make thoughtful choices. We have a tendency to consider speaking, reading, and writing to be automatic basics. Yet, today's computers are capable of reading printed text out loud to a person who cannot read, speaking words chosen by a person who cannot speak, and carrying out a wide variety of commands communicated without speech or even a knowledge of written English. For a person who needs aid in one of these areas of communication, it now becomes reasonable to ask which is more effective and which is more economical: teaching the person to improve communication function without technological aid or providing a device that reduces that demand on the person?

This distinction—improving vs. bypassing a skill—differentiates possible uses for the computer. In some cases, an educational aid is needed, a technique for helping the individual learn a new skill (e.g., reading) that he or she will later perform entirely unaided. In other cases, a permanent aid is needed—something that is capable of enhancing a person's communication by clarifying it, speeding it, or translating from the person's only available code (e.g., the limited movements and sounds of a nonspeaking, severely motorically impaired CP adult) into a more widely sharable code (e.g., speech or print). Which technique to employ must be thoughtfully answered throughout the course of each person's program.

While computers are still very far from listening in on free-running speech and recognizing anything intelligible in it, they may be able to provide enough supplementary information to a deaf individual to make speechreading really practical. Those speech features that are most difficult to detect by watching the lips of a speaker are relatively distinct to a computer and vice versa. Consider this example. Watch yourself in the mirror as you say *palm*, *mom*, and *bomb*. All three appear essentially the same to an individual who is speechreading. Yet, the *p*, *m*, and *b*, are acoustically distinct enough for the computer to tell them apart easily. By contrast, the initial phonemes in *palm*, *Tom*, and *calm*, are more difficult for a computer to distinguish acoustically. Nevertheless, we can spare the expense that would be needed to report those phonemes accurately because the three words look entirely distinct on the lips of the speaker.

A wearable speechreading aid using this principle was proposed over a decade ago. The aid looked precisely like a pair of eyeglasses. In an unobtrusive location, visible only to the wearer of the glasses, electronically generated symbols represented the acoustical information that the wearer could not detect easily by watching the mouth of the speaker. A decade ago, the idea was demonstrated feasible, but only with today's microtechnology has it become a very practical one.

A combination of two technologies relatively new even to the computer age has enabled the development of machines that read printed text out loud. Optical character recognition (OCR) is a complex statistical technique that analyzes the visual characteristics of a letter and decides what that letter must be. It is complex because letters appearing in various fonts, sizes, and print quality must be judged identical without sacrificing the ability to distinguish the subtle shapes that distinguish different letters. Just consider: guest, *guest,* **quest,** ***guest,*** quest, *guest,* quest, *quest,* **quest,** and ***quest.*** Once it is determined what letters are present, the task of converting that spelling into a pronunciation requires the application of a large number of rules. Letter-to-sound correspondences are complex in English, but orderly enough to be amenable to computer translation. The pronunciation of some words (e.g., *read* which may sound either like *red* or like *reed*, and *wind* which may rhyme either with *finned* or with *find*) must be determined entirely by understanding the context in which the word appears, something computers are not able to do in a general text. However, most words have unique pronunciations that can be determined by their spelling alone. Sophisticated text-to-speech systems have become quite good not only at pronouncing words but at coloring entire sentences with a reasonable approximation of normal prosody, the melody of speech that we sometimes refer to as vocal inflection.

"Simple" text-to-speech devices have been advertised at prices as low as $100, making it practical to use them much as one might use a printer on a computer. (See under Hardware in the Resources Section.) Text shipped to them is pronounced, rather than printed, and programs designed for non-readers who must receive their instructions orally, or non-speakers who need speech output to communicate with others, can be designed.

Criteria for Choosing Aids

Aids are not activities. They permit greater diversity in activity, as do eyeglasses, but they do not contain lessons, knowledge, or entertainment in themselves. The environment—what there is for the individual to do, to learn, to think about, to communicate about—remains the important focus. Still, aids should be considered wherever a handicap restricts an individual's autonomy and access.

There are those who feel it unadvisable to give an aid to anyone who, by struggling sufficiently, can get by without one. To say that an aid "can become a crutch," however, does not end any argument. Crutches help one get where one is going. When one is likely to go farther or faster with a crutch than without, it seems worthwhile to try the crutch. Educational activities cause growth, and are abandoned when they no longer provide significant rewards. Aids that are well suited to an individual allow that individual to engage in activities that will change him or her, and therefore will change his or her communication needs. Thus, the aids must be capable of changing, as well. They need to be easy to learn, easy to use, flexible, and expandable.

CONCLUSIONS

If we avoid using the computer as a surrogate teacher or drill-master, if we recognize that it cannot become a human-like conversational partner, what language role have we left for it? It can be just a machine, an engaging machine, designed to be sensitive to whatever skills the child *already* has for controlling it. A child gets immediate and concrete rewards for clever additions to the computer's repertoire, and also for additions to his or her own repertoire by being better able to make this phenomenal playmate play the right game. The emphasis of the interaction remains on the child's ability *to control*—precisely the emphasis it must have in order to help develop communication.

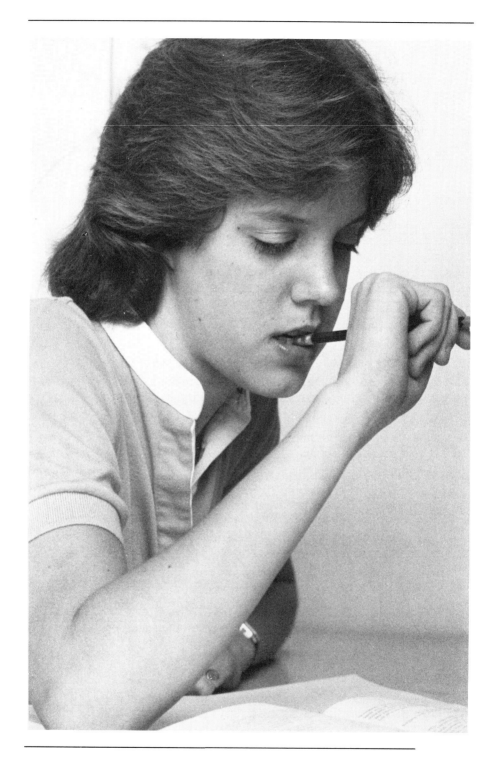

Mathematics: 4
A Case Study of Computers in the Curriculum

Mathematics is one of the curriculum areas in which the experience of special needs students might be enhanced through judicious use of the computer. Other areas—science, for instance, or the arts—have equal potential, but we have chosen mathematics as a case study of integration of the computer into the curriculum for several reasons. First, mathematics is not only a weak link in the education of many special needs students, but, in fact, it seems to *create* its own special needs population—a large number of students who fear, dislike, or are bored with mathematics. Second, mathematics is a prerequisite for an increasing number of college majors and vocational choices, so is a critical area for attention. Third, the view that mathematics is strictly sequential results in many students getting stuck, dropping out, being limited to a restricted set of mathematical experiences.

In the world of special education, mathematics, like art and music, is often ignored. Unlike the clear need for mastery of the tools of communication—written or oral, sound or symbol, biological or prosthetic—the necessity for skill and experience with mathematics is viewed, for the most part, as secondary. When problems of reading, speaking, or writing are paramount for a learning disabled, blind, or physically handicapped child, mathematics becomes a frill. It is included in the curriculum, but without enthusiasm. Functional mathematics is emphasized—keeping a checkbook, making change, telling time, measurement, and the bare bones of arithmetic. It is not unusual for a high percentage of any special needs population—visually impaired, hearing impaired, or physically handicapped, for example—to have poor mathematical skills; but those who provide special services for children in these categories receive most of their training in speech, language, reading, signing, and physical therapy. Few of those who teach children with learning problems receive much training in mathematics.

There may be many reasons for poor mathematical performance by special needs students: difficulty in holding a pencil, drawing a straight line, seeing a geometric diagram clearly, remembering arithmetic facts, or processing symbolic information. Inability to see the relationships of objects in space, to experience speed and distance, to use visual patterns to aid counting, or to hear verbal explanations all undermine the development of a body of knowledge about mathematical experience on which symbolic mathematics is based. Still, there is no *a priori* reason for a lower range of mathematical performance in many special needs populations, once the appropriate communication channels have been established and the lack of access to mathematical experience has been overcome. However, neither of these requirements is easy to address, and, even when physical or communication needs have been met successfully, mathematical deficits often persist.

In part, the absence of mathematics or the restriction of mathematics to a few applications in special education is a reflection of the place of mathematics in our society—a necessary evil, a subject to be mastered in order to be able to do elementary calculation, not to enrich the range of thought, creativity, and imagination. In part, it reflects lack of realization that mathematics has become the hidden entrance requirement to more and more fields. It is no longer just the sciences that require a solid mathematics background; mathematics is required for work in economics, psychology, business, management, and many other fields. A poor high school background in mathematics already drastically restricts vocational choice. Lucy Sells, a researcher concerned about the vocational access of women and minorities, has dubbed mathematics the "critical filter." Before even starting college, a high percentage of women have already lost the opportunity to pursue a large number of vocations because they do not meet the entrance requirement in mathematics for many majors. The same is true for special needs populations for whom little mathematics is required for high school graduation because other priorities seem more pressing.

Vocational access for those with special needs continues to be a critical issue. Suspicion, fear, and ignorance about their capabilities are barriers to equal vocational opportunity. However, the proliferation of computers in the workplace has the potential to open a range of new vocations to those who in the past had no chance to enter certain fields because they were homebound, could not speak clearly, could not hear, or could not write. The ability of the computer to store, retrieve, transform, graph, and print the user's ideas means that the communication and physical impediments to employment of the handicapped are slowly being removed. The computer can also be a tool in providing mathematical experience far beyond the functional mathematics which is no longer functional enough.

HOW MUCH MATHEMATICS IS ENOUGH?

What are the demands placed on the special educator today in mathematics? Much of the elementary and junior high school curriculum focuses on learning to add, subtract, multiply, and divide with whole numbers, fractions, and decimals. However, proficiency in arithmetic computation is no longer enough for mathematics literacy. Nevertheless, while some recent textbook revisions have paid more attention to applications and problem-solving, school mathematics in much of our educational system is still primarily focused on development of isolated arithmetic skills, divorced from the contexts in which they might be applied. The results of this emphasis alarm many educators. The National Assessment of Educational Progress study of mathematics achievement found that while many elementary and junior high age children can compute successfully when problems are presented in isolation, these same children may be unable to use their arithmetic skills appropriately. Results indicated that students know many mathematical procedures by rote but do not understand the concepts that the procedures represent. Performance at all age levels was extremely low on test questions that required problem solving beyond the level of application of memorized procedures.

In 1977, the National Council of Supervisors of Mathematics published a list of basic skills in mathematics, which has been used widely as a reference point for evaluating mathematics curriculum. Appropriate arithmetic ability is one of the basic skills listed. The other nine are: problem solving, applying mathematics to everyday situations, alertness to the reasonableness of results, estimation and approximation, geometry, measurement, use of charts, tables, and graphs, using mathematics to predict, and computer literacy.

The special educator is faced with a dilemma. Learning to calculate is not an adequate objective for school mathematics. Almost everyone can have access to a cheap calculator to assist basic computation, and this tool may be particularly appropriate for those who have struggled unsuccessfully for too long with these processes; but a calculator will not replace the ability to select appropriate input and operations, to make logical deductions, to interpret results, to assess the reasonableness of a result, to read a graph, or to build a geometric model of a mathematical situation. Yet, much teacher training for the special educator stresses arithmetic (insofar as mathematics is included at all), most curriculum materials cover primarily arithmetic and some functional skills, and most school curricula and standardized tests demand emphasis on calculation. When many special needs children are having quite enough trouble with the standard curriculum, how can we justify moving on to unfamiliar mathematical territory? Moving on, however, may be exactly what the situation demands.

The Myth of the Mathematical Hierarchy

> In fourth grade I was sick for three weeks—just when our class started long division. When I got back, I didn't understand what was going on, and I just never quite caught up after that.

> I loved algebra I, freshman year in high school, but I just couldn't manage geometry in sophomore year, so I never took algebra II.

Both of these statements reflect the general perception of mathematics, by both students and teachers, as strictly hierarchical. If you have not mastered level three, you are forever prevented from going on to level four and should not even consider level five, much less twelve or twenty-eight. Many children do get stuck at particular bits of mathematics—learning all the multiplication facts, borrowing, division of fractions, geometric proof—and remain stuck there. However, while it is true that, as in any field, learning more complex concepts is to some extent dependent on a foundation of prior learning, the rigid hierarchy of mathematical learning in schools does not reflect the actual relationships among mathematical ideas. Mathematics is a forest, crisscrossed by paths, not a mountain with a single trail. While some points in the forest are accessed most easily by a particular path, there are many trails that lead to scenic spots and many routes that interconnect unexpectedly. Depending on what you are looking for, some paths are better than others—some lead to water, some to wild herbs, some to caves, some to rare flowers. Some require skill at climbing, others sharp eyesight, others a good memory for landmarks.

As an illustration of this diversity, consider the following problems. If you wish, try each one. A discussion of the problems can be found at the end of the chapter.

1. I just flipped a dime four times. Each time it came up heads. What is the probability it will come up heads on the fifth flip?
2. Estimate the answer to: 12/13 + 7/8.
3. Mr. Jones put a rectangular fence all the way around his rectangular garden. The garden is ten feet long and six feet wide. How many feet of fencing did he use?
4. Why is the graph in Fig. 4.1 misleading?
5. About how much change will you have left from ten dollars if you pay a lunch bill of $6.28 and leave a 15 percent tip?
6. What's wrong with this advice to runners: Before you attempt to run in high temperatures, you should be able to run a mile in eight minutes, or 1.6 kilometers in about one minute.
7. What solid figure will you get if you rotate the triangle in Fig. 4.2 around the side AB?
8. What is the next number in each sequence?
   2 4 6 —   1 2 4 —   8 5 4 9 1 7 —

Problems 1, 2, 4, and 7 require no computation. Problems 3 and 8 require some computation, but success in solving them is determined primarily by how well the structure of the problem is understood, not by computational skill. Problems 5 and 6 may require a good deal of arithmetic, but estimation, choice of strategies, and familiarity with the context are more important

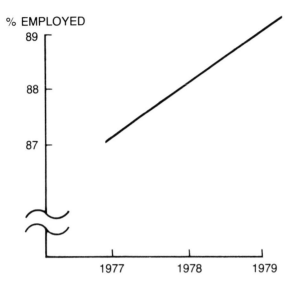

Fig. 4.1

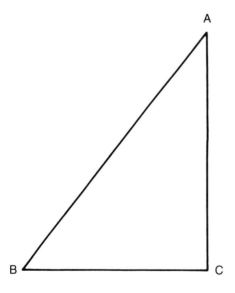

Fig. 4.2

factors. It would be difficult to order these problems in terms of difficulty or by level of skill required. Experience with the type of problem, ability to visualize spatial relationships, understanding numerical relationships, ability to extrapolate, interpret, estimate—all these affect one's ability to solve these problems as much, or more than, arithmetic skill.

So what of the mathematical hierarchy for special needs students? Should inability to do long division prevent a child from grappling with probability, estimation, graphing, or geometry? Because it comes first in the curriculum, we assume that learning the multiplication tables must precede finding the area of a parallelogram. It is not so much that we think learning 6 × 7 is a *prerequisite* for this activity but that we think inability to memorize multiplication tables (or do borrowing or add fractions) is *indicative* of a level of mental ability which prevents the student from tackling "higher" mathematics. Yet we know that mathematicians may be good at only particular pieces of mathematics and, can, in fact, be poor at "lower level" skills, such as computation. More important, the skills needed for different kinds of mathematical problem solving may be unrelated. In a detailed study of children with unusual talent in mathematics, the Russian psychologist Kruteskii found children who were exceptional in either visual or analytic approaches to mathematics and relatively poor in the other. The implication for evaluation of mathematical ability and for the planning of the mathematical education of special needs children is significant. Students with impairments that prevent success in some kinds of arithmetic may not only find other areas of mathematics more accessible but may also reveal special strengths previously untapped.

This is all very well. But how is a classroom teacher to find the time, materials, resources, and support to provide individualization, give students the feedback which will allow them some control over their own learning process, explore the variety of mathematical experience with each child, and yet cover the expected curriculum—not only mathematics, but all the rest?

DRILL AND PRACTICE

Use of computers in the mathematics education of special needs children dates back at least to 1970 when the Office of Education funded Patrick Suppes' project at Stanford University. The computer-assisted mathematics instruction part of the project was initiated at the California School for the Deaf in Berkeley in 1970 and at the Florida School for the Deaf in 1971. These schools and many other schools for the hearing impaired have adopted drill-and-practice computer programs. Drill and practice in mathematics has also been used with other special needs populations, including the physically handicapped, retarded, learning disabled, and those with severe language handicaps. Such systems allow some selection by the teacher of difficulty level, sequence, and content. In addition, some schools have adopted or developed authoring systems, which allow teachers to design individual drill-and-practice lessons for one or several children.

A typical drill-and-practice mathematics lesson in its most unembellished form consists of the presentation of one problem at a time followed by the student's response and then a message, e.g., "Correct!" or "No, try again." This kind of instruction is viewed primarily as reinforcement and practice of previously introduced classroom work. The advantages of such software most often cited by teachers of special needs populations include: 1) it is easy to use; 2) it allows repetition and practice in small increments; 3) it provides the student with immediate feedback; 4) it can keep records of students' answers; 5) it allows generation of an endless supply of practice problems, tailored to the level of the individual child; 6) it frees teacher time to provide more individual attention to students.

Reports of results of using drill-and-practice programs with special needs populations have ranged from "no effect" to "marginally effective" to "significantly beneficial." Many students appear to enjoy the privacy and predictability of their interaction with the computer in this mode and use of arcade-like graphics and speed can certainly make drill and practice programs, such as the Arcademic Skill Builders, attractive. However, it is important to realize that drill-and-practice computer-assisted instruction (CAI) taps only a small part of the potential computer use in mathematics education. It is also important to recognize that at a time when the mathematics curriculum is expanding and when the need is to improve problem-solving skills, drill-and-practice software systems emphasize those elements in the curriculum that are already thoroughly dealt with in traditional textbook curriculum materials. The results of a study of drill-and-practice CAI, conducted by the Educational Testing Service and the Los Angeles Unified School District, indicated that the use of computers in this way could improve arithmetic skills. However, in grades 3–6 the effects of the computer instruction on concepts and applications (as measured by scores on a standardized achievement test) were very small. The ETS report argues that these results are inconclusive because reading is a factor in the concept and applications sections of the test. An alternative explanation is that emphasis on calculation improves only calculation, as might be expected.

The child or adult with special needs should not be cheated out of the endless variety within mathematics. Some paths through the mathematical forest may be inaccessible, but there are thousands of paths, each with its own scenery, shortcuts, and unexpected connections to other routes. Focusing only on weaknesses in arithmetic or on practical mathematics seriously restricts access to mathematics and may result in the atrophying of a gift which a child might discover by following some less travelled road.

THE WOODS ARE LOVELY, DARK, AND DEEP: BEYOND DRILL AND PRACTICE

Computer assisted instruction in mathematics can go beyond straight drill into more creative concept and application development. Carefully designed computer software can combine graphics, text, and animation in an interactive context, allowing students to play with new ideas and build up new structures of mathematical understanding, as slowly or as quickly as an individual user may wish, in privacy or in cooperation with other students.

Beyond drill and practice, there are three potential uses of the computer in the mathematical education of the special needs child. First, the computer can provide educational games that use dynamic, graphic, interactive models of how mathematics works. Second, the computer makes available mathematical tools for organizing, recording, and visualizing mathematical information. Third, and perhaps most significant in the long run, is the creation of computer microworlds that allow children (and adults) to engage in mathematical play and exploration in realms not otherwise easily accessible to them.

Learning Problems in Mathematics

Consider again the problems on page 91. If you tried these problems, you probably found some to be easy, some difficult. Some were already familiar types of problems. For others, you had knowledge or experience that made access to the problem straightforward. Others may have baffled you, or you may have thought you had found a solution but were mistaken. If you were unable to do a problem, why was this? Among the possible explanations are the following:

1. A physical handicap prevented you from seeing or recording information.
2. A learning problem prevented you from visualizing the problem, processing information, storing information, concentrating, or organizing.
3. You "just can't do math." (Or, similarly, you "just can't do problems with fractions," or whatever.)
4. You never had experience with the particular problem type.
5. You are unfamiliar with a context or model within which the information in the problem would make sense.
6. Something about the way in which the problem was presented distracted you from the essence of the problem—which, in fact, you understood well once you read the explanation.

Special needs in education arise from all of these sources. Severe, specific learning problems may be entirely responsible for a small percentage of school mathematics failure. However, many students experience failure because of a complex interaction among some or all of the following elements: mild learning problems; physical, auditory, visual, or communication needs; inexperience with mathematics; impoverished or boring curriculum; fear and anxiety; belief that mathematics is unimportant and unrelated to real life; belief that school mathematics is governed by a set of arbitrary and unfathomable rules, created by unknown authorities and urelated to common sense.

Let us consider a few special needs children, why they have difficulty approaching mathematics, and how the computer might enhance their mathematical education. Although none of these descriptions is based strictly on an individual child, all are drawn either singly or as a composite from real children in a variety of school settings.

Educational Games

Mason is a 10-year-old child in an intermediate classroom for the hearing impaired. He is mainstreamed for part of the day into a regular fourth-grade class, and children from the same fourth-grade classroom spend some time each week in his class. All Mason's skills are about a year and a half below grade level, which is not unusual for those in his classroom. He does fairly competent work in addition, subtraction, and simple multiplication, but he seems unable to comprehend multi-step processes such as double-digit multiplication and long division. Recently, his teacher reached an impasse with Mason in attempts to work on fractions. He can recognize and write the symbols for 1/3, 1/2, 1/4, 1/6, etc., and color in corresponding pieces of pictures, but he has been unable to move further into work with equivalent fractions. This work has become so frustrating for him that when he sees a page with fractions on it, his face becomes blank, he turns away, refuses eye contact, and cannot be turned toward the work by persuasion, firm insistence, or even anger. This behavior has appeared before, and neither his parents nor the school has found a successful way of dealing with it: either the topic will have to be re-introduced in a new, disguised form or dropped altogether. Since this total resistance only happens occasionally, his teacher is willing to ignore most of these instances. She needs all the energy she has to deal with the needs of each member of her class for individual attention in both academic and social-emotional areas, and learning fractions seems like a relatively minor issue or, at least, not worth the amount of time and anguish it looks like it might require at this stage in Mason's schooling.

It is clear that Mason needs individualized instruction, but he needs more: he needs a way into fractions which will allow him to gain understanding slowly, with privacy, adequate feedback, and control of his own learning. Like many children with special needs, ranging from hearing or speech deficits to learning disabilities and physical handicaps, Mason has met failure often. When a new expectation arises through material which appears strange, mysterious, or out of his control, he turns off. Mason might enjoy one of several computer activities that deal with fraction concepts. Unlike drill-and-practice programs, these programs help children to become familiar with a visual model of fractions, a model which allows them to develop mental representations of the relationships of fractional and whole numbers. At the same time, they are engaged in an interactive game format that is fun and private.

A computer game called *Darts* presents the user with a vertical number line on which a few points are labelled. Several balloons are attached to

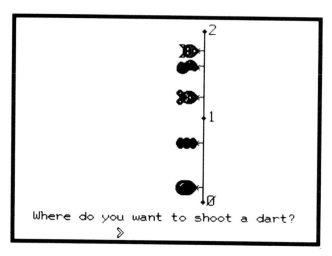

Fig. 4.3

unlabelled points on the line (see Fig. 4.3). The student attempts to pop the balloons by typing a number on the keyboard to correspond to the guessed location of a balloon. A dart moves from the side of the screen, hitting the numberline at the point specified by the typed number (Figs. 4.4, 4.5). A dart can hit any point on the balloon to pop it, so there is actually a range of correct answers for each balloon; the option of large or small balloons varies the difficulty. An important feature of this game is that there is no penalty for an incorrect answer. A dart which does not pop a balloon simply does not pop a balloon, but, in addition, it provides more information by giving an additional point on the line (Fig. 4.4). This game provides a visual, dynamic model that will make more sense to many children than numbers in isolation. They receive immediate visual feedback about their selections; they have control over their choices and receive only objective information, not negative and empty feedback like, "Incorrect, try again." Children playing *Darts* find that there is more than one answer which works. One child may choose 1 5/12, another 1 6/12, another 1 1/2, another 17/12; all of these may hit the same balloon.

Another game links fractions with probability. *The Jar Game*, one of a series of mathematical computer games developed by William Kraus, presents two jars, each filled with a mixture of green and gold pieces. The user attempts to select the jar which contains more gold than green.

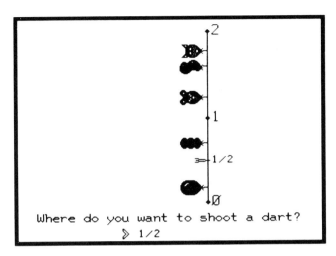

Fig. 4.4

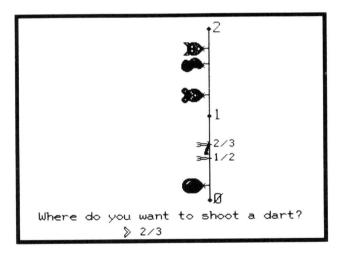

Fig. 4.5

Once the selection is made, a fly buzzes from one piece to another in the selected jar and lands on one piece. If it lands on a gold piece, the user scores a point; otherwise the computer scores a point. If the user chooses the jar with the larger proportion of gold pieces (not the larger *amount*), the *chance* of beating the computer is better. This game provides a model of fractions which is quite different from the number line model; it combines experience with probability and preparation for ordering fractions.

A third fraction activity is part of the PLATO elementary computer mathematics program. *Paintings Library* requires the student to color in a particular fraction of a rectangle in any way he or she chooses. By touching the screen, the user can draw lines and color areas of the box. If Mason chooses to save his painting in the painting library, other students can see his work. Teachers who have used this activity note that students rarely want to imitate another child's painting. Instead, they use others' work to stimulate new ideas of their own. For instance, if one child paints her initials, using half of the rectangle, other students often try their own initials. Some examples of the paintings children produce appear in Figs. 4.6, 4.7, 4.8, and 4.9. This last activity, in particular, shows the potential of good computer software to allow children to play with mathematical concepts; students begin to share ideas about mathematics in the way they might share stories they have written.

All three of these activities enable children to use fractions in a context that makes sense. With access to a variety of models, children can begin

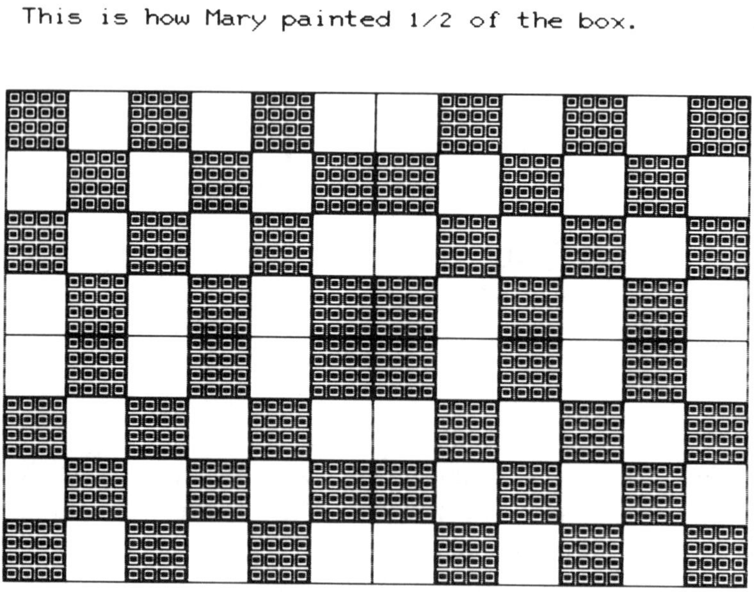

This is how Mary painted 1/2 of the box.

Fig. 4.6

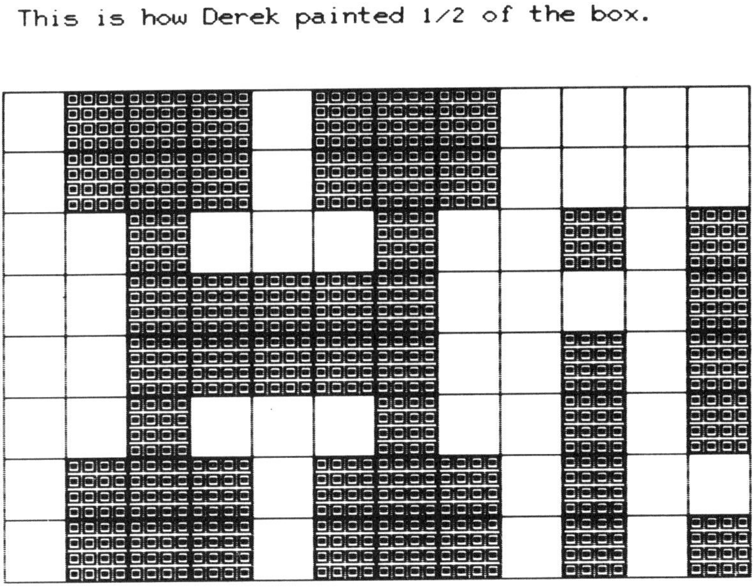

The Woods Are Lovely, Dark, and Deep: Beyond Drill and Practice **101**

This is how Derek painted 1/2 of the box.

Fig. 4.7

This is how Frederic painted 1/2 of the box.

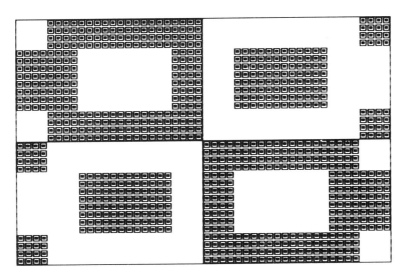

Fig. 4.8

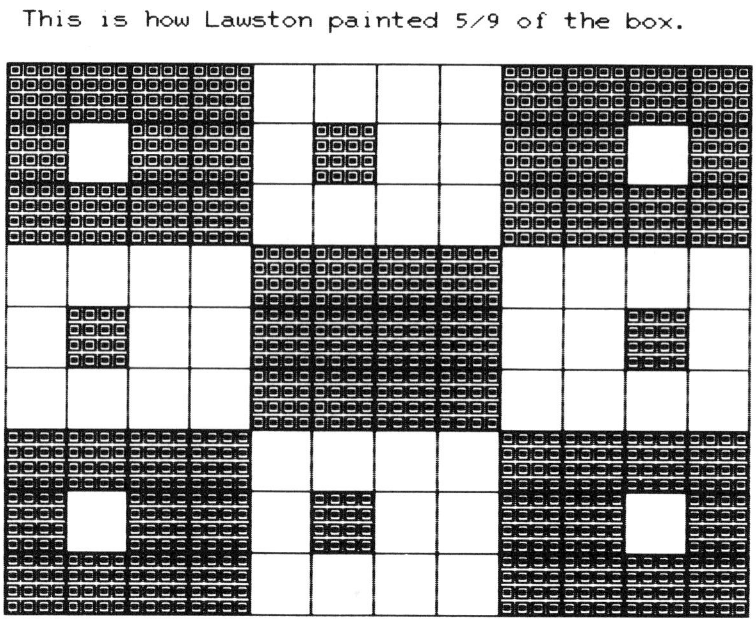

This is how Lawston painted 5/9 of the box.

Fig. 4.9

to build, change, and expand their own internal models as they see what fractions denote equivalent amounts, how a fractional amount can be represented in many different ways, how a fraction may denote a position (1/3 mile from the intersection) or a quantity (1/3 cup of flour). These computer activities provide individualization, privacy and control for the child, and variety. They do not provide all the kinds of experiences with fractions that children should have. For instance, another PLATO activity, *Make-a-Monster*, which has children follow a recipe (2/7 cup icky liquid, 1/4 cup creepy cream, etc.) to produce a suitable monster formula, may be fun, but it is no substitute for mixing up a real batch of chocolate chip cookies.

Another student, a ninth grader we will call Arlie, provides a second example of appropriate use of high quality educational games. Arlie is a self-taught artist with a sophisticated sense of perspective and proportion. She spends her spare time illustrating stories. She has difficulties in school, mostly because she is inattentive and disorganized; some have called her hyperactive. In fourth grade she was tested by the school psychologist whose report indicated that she was achieving considerably below her potential. Her end-of-year reports consistently contain phrases such as "short attention span," "low frustration tolerance," "poor concentration," "no initiative or perseverance," and sometimes "disrespectful" or "troublemaker." She manages to produce adequate reading and English assignments, although her spelling and punctuation are poor, but her work in mathematics is always borderline. She is able to memorize some rules and apply them, but more often she half-remembers a process, applies the right rule in the wrong situation, or mixes up the sequence of steps. However, during a unit on geometry in eighth grade, her grades in mathematics temporarily soared. While she could never remember the formula for interior angles of a polygon, she almost always answered questions about length and angle correctly. Here are two examples of her work:

> How many degrees in one interior angle of this equilateral figure (see Fig. 4.10)?
>
> 60 degrees
>
> How did you figure that out?
>
> Oh, I could just tell it was 60.
>
> What is the area of the circle (Fig. 4.11)? a) 44 b) 154 c) 49 d) 196

Arlie also answered the second question correctly, but this time, although she did not use the formula for the area of a circle, she did have an explanation for her answer: "Well, if it was triangles, they'd be about 25

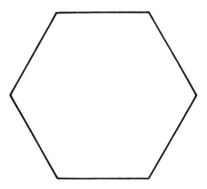

Fig. 4.10

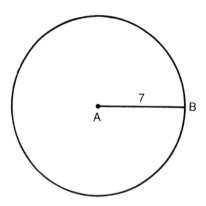

Fig. 4.11

each—that's 100—but it's more than that because of the extra, and 196 is too much because that would be about the whole square." Can you follow her reasoning?

Arlie is a child who does poorly in much school mathematics, but may actually have a flair for problem solving using her particular style, perhaps even in the areas of mathematics in which she presently does poorly. However, when asked about mathematics by the guidance counselor this year, she is sure (and her counselor agrees) that she wants to take the minimum high school requirements. After all, she will never use mathematics as a graphics designer. Arlie has this to say about mathematics: "The girls don't really care about math. They are not very interested because it has so many rules. You have to do this and you have to do that and you can't do this and you can't do that. You hate the rules and you want to be doing something else so you don't remember them. When I draw, it doesn't have to be real; it can be a part of my imagination. In math you can't do that because of the rules; you can't go out and tell a fairy tale."

Arlie is having extreme difficulty in her algebra course, although it is a slower-paced course combined with continued work in general mathematics. Her teacher hopes that a computer game called *Green Globs* will allow her to use her understanding of geometric relationships to move toward improved understanding of algebraic equations.

In *Green Globs* the students are presented with coordinate axes and a number of green globs which they must try to hit. They attempt to hit

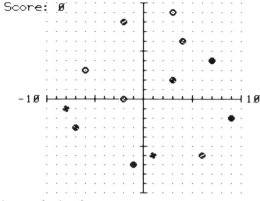

Fig. 4.12

as many globs as possible by writing equations which the computer obligingly graphs (Figs. 4.12, 4.13). Users begin to see the graphic effect of changing an equation from $y = x$ to $y = x + 3$ or $y = 3x$ or even $y = x^2$ and to learn what it is sensible to expect about the behavior of algebraic equations. For children like Arlie, who get lost in the manipulation of symbols, such an activity helps acquisition of some important generalizations about algebra, which their difficulty with symbols might otherwise obscure.

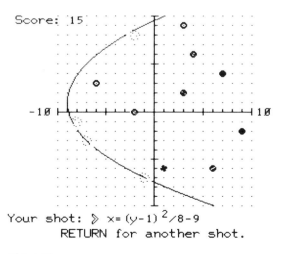

Fig. 4.13

With careful selection, teachers can find sound computer-based activities for remediation and new learning. The best of these have several advantages: children enjoy these activities; usually the level of difficulty can be varied; they provide nonthreatening, clear, useable feedback; they provide a variety of models of mathematical ideas, more than may be readily available to the teacher or student through textbooks or other materials. For the special needs student who has not learned easily using conventional methods, this access to new models is especially important.

The Computer as a Mathematical Tool

Students who have little difficulty understanding mathematics may nevertheless be seriously impaired in their ability to do mathematics because they cannot organize and record information. For them, the computer can act as a prosthetic for doing mathematics. Children who cannot easily manipulate a pencil, paper, or other mathematical materials can make responses and receive feedback in the modes which work best for them. Modifications to the computer to expand the range of students who can use it may include touch-sensitive screens, light pens, Braille keyboards, and devices that translate text to touch-sensitive or voice-synthesized output. The computer may also be used as a scratch-pad for mathematical problem solving, allowing handicapped students to store intermediate results and other information. Software packages that do graphing or calculation are particularly useful for the older student who has good problem-solving skills but needs assistance in recording and/or organizing information.

Karen is a senior in high school and a quadraplegic. Because her speech is also severely impaired, her interaction with the world had always been slow and usually mediated by a scribe. When she was in tenth grade, her school acquired a microcomputer, including a word processing system which Karen learned to operate using a headstick to push the keys. While her ability to write improved rapidly as she mastered the system, her mathematics continued to lag far behind an age-appropriate level. Her teachers did not know whether this lag could be attributed primarily to lack of experience, lack of understanding, or inability to record while solving problems. Karen could do most elementary calculation in her head, but she appeared frustrated when attempting to solve multi-step problems, to use more than two numbers at a time, or to calculate with numbers having more than two or three digits. One morning Karen's teacher noticed that she was using the word processing system to do her mathematics homework, recording intermediate steps and "jotting down" bits of information. Since a word processor is hardly an ideal tool for doing mathematics, the school began a search for a "mathematical word processor."

They soon found a number of useful tool programs. While not designed specifically for physically handicapped students, these programs matched Karen's needs, and it became apparent that a large measure of what had held her back in mathematics was simply an inability to keep track of the information her mind was capable of handling.

SemCalc, short for "semantic calculator," enabled Karen to record and manipulate both quantities (e.g., 23) and their referents (e.g., miles per gallon). The program first presents a notepad with two columns labelled *How Many?* and *What?* on which the student can record all possibly relevant numbers and their units. For instance, the student might record:

	HOW MANY?	WHAT?
A	218	DOLLARS/WEEK
B	52	WEEKS/YEAR
C	5	DAYS/WEEK
D	40	HOURS/WEEK

With this information recorded, the student can direct the computer to do any calculation with any of the items. for instance, she might choose to multiply quantity A and quantity B. The program lets the user know that the units of the answer for that calculation will be DOLLARS/YEAR. If this is the result she is after, she can have the program proceed with this calculation. If not—for instance, if she meant to find out how many dollars are earned per working day—she can request a different calculation. The program allows users to record information and to focus on the reasonableness of their results in relationship to the original problem. It does not tell the user whether the answer is right or wrong or even if it is reasonable. These are judgments the user must make herself. But it does provide the user with an organized electronic scratch-pad which makes it easier to make those judgments.

Other useful software for Karen was soon discovered. With a program called *Speed Up Your Algebra*, Karen can solve algebraic equations by telling the computer which step to take next. For example, she can type

S 30 to indicate that the computer should subtract 30 from both sides of the equation. With another program, *Plot,* she can see any equation she specifies graphed on the screen.

As Karen became more proficient and self-assured in her use of mathematics, her teacher introduced her to *VisiCalc*. *VisiCalc* presents a large matrix, 260 rows by 70 columns, in which the user can enter quantities, labels for the quantities, or formulas. For instance, row 1 might contain population figures for 100 major cities in 1973, and row 2 might contain population figures for the same cities in 1983. The user could direct the program to calculate the change in population from 1973 to 1983 and place the results in row 3, the percentage of change in row 4, and the projected change by 1993 in row 5 by entering the three appropriate formulas. Karen is now using *VisiCalc* to calculate elementary statistics for a long-term science project involving growth of bacteria.

Playing with Mathematics

When was the last time you played with language? In the past week, have you told a joke, enjoyed a conversation, written a poem, been carried into an imaginary world through something you read, been stimulated by listening to someone talk about a new idea? When was the last time you played with mathematics? In the past week, have you solved an interesting numerical problem, discovered a novel geometric tiling pattern, estimated the number of insects attracted to your window during an evening, analyzed a misleading advertising survey, challenged a friend to a game of strategy? Perhaps you have done some of these, but have not thought of them as mathematical. Whereas most of us use and enjoy some aspect of language routinely, we are hard-pressed to come up with experiences of mathematics in our lives beyond the minimum we need to function—paying a bill, estimating miles per gallon for the family car, and so forth. Play with mathematics is not a usual component of our lives—nor of our educational system.

For most of us, time to explore mathematics, to create and solve our own problems, is almost nonexistent. For children with special needs, who often must concentrate primarily on what they cannot do well, there is no time to play with mathematics. Yet, as David Hawkins has pointed out, children learn a great deal from "messing about" with scientific and mathematical ideas. However, it is not so easy to arrange productive, exploratory mathematical experiences in the classroom.

Perhaps the most important contribution the computer can make to mathematics education for special needs students is to provide access to mathematical worlds in a way that has seldom been possible for special needs populations. An important proponent of this use of the computer is Seymour Papert, who with his colleagues created a computer language called Logo, which provides such a mathematical "microworld." Logo begins with what children already know about moving themselves through space. They use this knowledge to direct the Logo turtle, which may be a robot turtle that moves around on the floor, or a small triangle that moves around on the computer's graphics screen. The turtle can be commanded to move forward or back, to turn right or left, to draw as it moves or to leave no trace. The power and flexibility of the Logo language can be seen in the variety of its uses by children and adults: young children may use it as a simple drawing tool or as a setting for dramatic play; older students and adults may use it to explore sophisticated mathematical ideas and programming techniques.

A young child or a beginner at any age might, at first, simply move the turtle around the screen, creating unplanned lines and shapes. Often the lines and shapes suggest a particular pattern or picture leading children to define their own problems: "that looks like a bird—I think I'll make wings on it," "I want this to be the same on both sides," "I want to close up this shape." In attempting to achieve particular effects, students become engaged with estimation, matching, rotation, and prediction. Figs. 4.14 and 4.15 show two pictures created by a 7-year-old and an 8-year-old. At a more advanced level, students might explore the characteristics of circles, polygons, trigonometric functions, spirals such as those in Figs. 4.16 and 4.17, or complex recursive patterns such as the one in Fig. 4.18. Logo's structure encourages children to break down complex problems into manageable parts, solving each separately until the entire problem is completed.

Kenji is a 12-year-old child with severe learning problems. He is unable to do most computation required at his grade level. Minute differences in the format of a familiar type of problem (for instance, a subtraction problem presented horizontally rather than vertically) are enough to defeat him completely. He reads with difficulty and is an extremely poor speller,

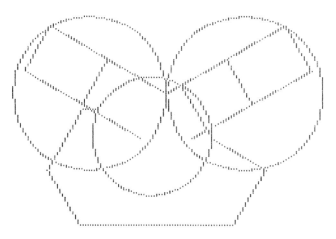

Fig. 4.14

continuing to misspell common sight words such as "are" and "there." When he first began work on the computer, Kenji made so many typing mistakes that his teacher thought he would be unable to use the keyboard. Not only did he have difficulty finding the letter he wanted, but once he found it, he was so impatient to go on to the next letter that he often took his eyes away from the first letter too soon and hit the wrong key.

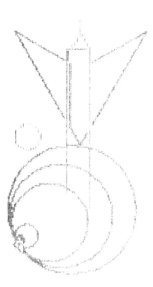

Fig. 4.15

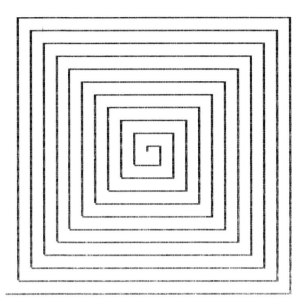

Fig. 4.16

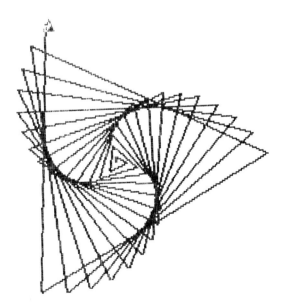

Fig. 4.17

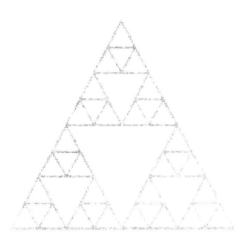

Fig. 4.18

However, Kenji found his ability to control the Logo screen turtle so appealing that he persevered despite trouble with typing ("Why can't they put these letters in order?" he said one day in frustration). To his teacher's surprise, Kenji, who does not recognize the unreasonableness of many of his answers in his school mathematics work, showed a strong ability to use numbers sensibly in his Logo work. Although he knew nothing about angle measurement, Kenji "invented" the 90-degree angle in his first session and figured out how to make a circle and modify its dimensions in the second. When he works at the computer, Kenji is engaged, excited, attentive, and is, perhaps for the first time, using his errors as a source of information about what to do next rather than as proof of failure. In addition, his projects have led him to play with some important mathematical ideas, including estimation of length and angle, the relationship between interior and exterior angles, and the relationship between polygons and circles. He is even beginning to use computation in this context without so much fear. To center a door of length 30 in a drawing of a house of length 150, he accurately sequenced and calculated the two-step process: divide 150 by 2, subtract half of 30 to find how far from the corner the door should be placed (point A in Fig. 4.19).

Kenji's experience with Logo provides the briefest of glimpses into what the experience of mathematics could be like for many children— an opportunity to discover that piece of mathematics which they might love: geometry, topology, probability, statistics, number theory, algebra,

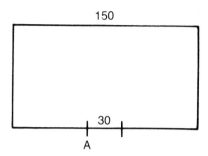

Fig. 4.19

logic, applications to art or science. For a more extensive discussion of Logo, as well as many of the other uses of the computer in mathematics education, see *Computers in Teaching Mathematics* (Addison-Wesley). The next section offers a view of how such a mathematical education might begin to harness the potential of the computer, using it for a range of activities, from limited-goal instruction to simulation to mathematical play.

A WALK THROUGH THE FOREST

Almost all of the computer activities currently used in special needs classrooms are in the form of computer assisted instruction, either drill and practice or other limited goal educational games. Computer assisted learning activities range from the dull and silly to the inventive and useful. But such computer assisted curriculum only begins to tap the potential of the computer for special needs students. The computer offers new paths into the mathematical forest—perhaps even some paths unknown to the teacher—through play, exploration, problem solving, and new learning. What follows is fiction, but all the software is quite real and currently available.

Imagine a semi-urban community of 70,000 which, within its public schools, serves a wide range of special needs students, from those with mild learning problems to those with severe physical handicaps. In one of the schools (grades kindergarten through eight) there is one classroom for each grade and a separate class for hearing impaired children. Children with physical handicaps and mild to moderate learning or emotional problems are mainstreamed into the regular classrooms. The school has four microcomputers: three are located in the mathematics laboratory, and the fourth, on a rolling cart, is reserved by individual classroom teachers for a week at a time. Each classroom is assigned mathematics lab time for three periods each week. Most teachers send half their class to the lab during one of their periods, the other half during the second period, and use the third for children who need extra tutoring or are finishing projects.

During second period on Monday morning, ten third graders, a bit rambunctious yet from the rainy weekend, come to the mathematics lab. The school has made a conscious decision to keep these groups heterogeneous, so the abilities and needs of these children vary considerably. Two children, Fania and David, are assigned to some work with two-digit addition; Terry and Denise work on a game of sorting and classifying with attribute materials. The other six children use the microcomputers.

Jimmy, a poor reader, often described as hyperactive, uses the Logo language on one of the computers. He still has difficulty finding the keys on the keyboard, even after many sessions, but he has partially solved this problem with his teacher by sticking colored circles on the keys he uses most often. Despite his problems with the keyboard, Jimmy has learned many of the commands that direct the Logo screen turtle. Jimmy shows a flair for accurate estimation of length and angle, and he has begun to incorporate careful forethought and planning into his Logo work. He likes to plan pictures with several parts, working on each part separately until he is satisfied with it, then combining his small programs into the final large one. Currently Jimmy is working on a traffic light (Fig. 4.20), which he plans to add to a picture he has already created (Fig. 4.21). The frenetic activity with which he entered the room has all but disappeared, although he continues to hear (and reply to) everything everyone else says in other parts of the room while he works.

Fig. 4.20

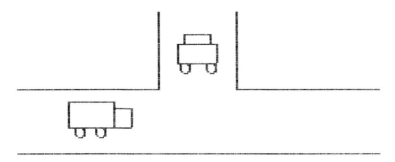

Fig. 4.21

At the second microcomputer are Carmina and Franny, immersed in a microworld of attribute activities called *Gertrude's Puzzles*. They are attempting to arrange nine puzzle pieces (three diamonds, three triangles, three hexagons) in a 3 × 3 arrangement so that no row or column has more than one piece of any particular shape or color. They move the pieces on the screen by using four keys with arrows (up, down, right, left) and the space bar for picking up and dropping individual pieces. Franny, who has only partial use of one hand, can hold a pencil and push one key at a time; because of her limited hand use, she is unable to manage games with the physical attribute blocks that Terry and Denise are using. Franny has become particularly intrigued by the facility this program offers which allows the users to create their own puzzle piece shapes. By moving to the "Shape Edit Room," Franny can begin with one of the given shapes, which is enlarged on a grid (Fig. 4.22), and by using two keys (E for erasing, spacebar for filling in), she can create as intricate a shape as she wishes (for example, Fig. 4.23). When she is finished, *all* the objects shaped like the original piece will appear in the form of her new shape.

The remaining three children, Claude, Corey, and Steven, are playing a simulated game of golf on the third microcomputer. The game involves angle estimation (using standard degrees) and length estimation (using arbitrary units). The length estimation fits well into the measurement activities they have been doing in their classroom with standard (inches, centimeters) and nonstandard (paper clips, shoes) units. The mathematics lab teacher had not thought of using this particular activity with third

116 *Mathematics: A Case Study of Computers in the Curriculum*

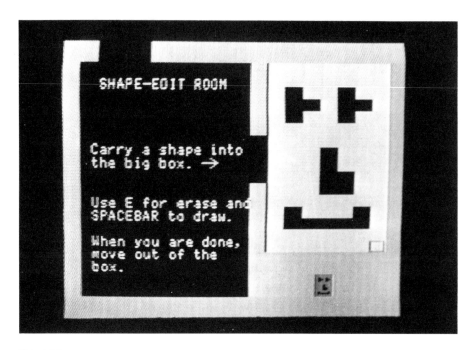

Fig. 4.22

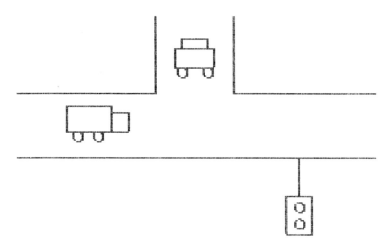

Fig. 4.23

graders because of the angle estimation involved, but Corey saw some older children using it and asked to try. To send the golf ball to the hole, the player chooses first an angle, then a distance. A diagram on the screen helps the player choose an angle, and if the player types "H" (for "help"), the diagram is placed directly over his or her ball to make the choice even clearer. With these aids, the third graders have begun to become quite proficient at choosing a sensible angle. For each game, the length of one unit changes so that the children have to change their basis for estimating distance from game to game. Corey, who is hearing impaired, loves this game; he is good at it, feels in control of the situation, and he and his non-hearing-impaired classmates relax and relate to each other easily as the game progresses.

By the end of the period, a dramatic cry of "oh no!" followed by giggles from Jimmy has caught Carmina's attention and she has moved over to help Jimmy. He has finished his traffic light, but has not quite managed to integrate it into his previous drawing. They settle down to figure out how to fix it. Franny, left to her own devices, becomes absorbed in creating a new attribute piece. Corey, Claude, and Steven finish their game and move on to complete work in their individual folders. Fania, David, Denise, and Terry move to the computer now left free by the golfers to continue a simulation game they started last week called *Lemonade*, in which teams run lemonade stands and make decisions about how much lemonade to make, what to charge, and how much to spend on advertising. Fania and David's work with two-digit addition finds an immediate application here.

Most of these children come to the mathematics lab once a week. The number of computer hours available per child is hardly ideal for this school population of 225. Because teachers and parents are convinced of the opportunities and variety the computer offers, the PTA is attempting to raise money for an additional computer, and a group of teachers is working on a grant to obtain another.

What most impresses the visitor to this school is that children have begun to share mathematical experiences and ideas with each other and with teachers and parents. Because they encounter such a variety of activities during mathematics lab time, they think of mathematics more broadly and tend less to classify themselves as "bad at math" solely on the basis of their ability to compute. While computation is worked on seriously, children participate in mathematics in other ways as well, so that individual strengths have a chance to flourish. Computers are an integral part of this endeavor.

FINALLY, GUIDEPOSTS AND WARNINGS

The computer is a powerful tool for doing mathematics but it is not a panacea. While surveying the possibilities the computer offers, keep in mind five warnings. First, computers in mathematics education may be used narrowly; their potential for development of concepts and understanding may be ignored.

Second, computer use can be inappropriate or even silly: one program for hearing-impaired students claims to teach use of a ruler by giving written directions on the screen; another remedial system used with physically handicapped children delivers drill work at 1500 skill levels (more than one level per day for eight school calendar years).

Third, computer-student interaction can be unimaginative, too difficult, not challenging enough, mystifying, or just plain dull. Therefore, the role of the teacher remains critical in selecting, matching, and integrating computer experience into the mathematics curriculum. The computer not only does not replace the teacher, it demands the same hard work of observing, questioning, leading, and supporting in a new and unfamiliar domain.

Fourth, use of the computer, like use of a library, should be partly controlled by the child. While adults can be supportive in helping children choose books, we hardly want to restrict their selection to books we have read. Similarly, if enough good software is available and children begin to make selections based on their own interests and peer recommendations, they may find themselves engaged in mathematics unknown to their teachers. Teachers must be brave enough to allow children to play with unfamiliar mathematics.

Finally, use of the computer is not *the* solution any more than is any other educational tool. It is a good, partial solution to some of the mathematics education needs of many children. Educators, parents, and administrators must take an aggressive role in deciding which educational software, combined with which other mathematical tools, best serve the needs of particular children. Teachers' knowledge of how children learn and of the scope of mathematical content is even more critical than in the past, as computer program designers, who may know neither about children nor about mathematics, flood the market with computer activities. Students can be guided along paths in the mathematical forest only by someone who both knows something about the abilities and interests of the explorer and has a general map of the forest and its possibilities. However, gaps in teacher knowledge of mathematics should not be perpetuated and transmitted to students by allowing student access only to what an individual teacher understands. The teacher's map does not have to include all trails, and the good teacher will be alert to the student's own discoveries. Belief in a mathematical hierarchy may have to be modified, perhaps abandoned.

The woods are lovely, dark, and deep, indeed. And we have many promises to keep in giving all children an adequate mathematical education.

DISCUSSION OF PROBLEMS

1. The probability of a coin coming up heads, assuming a balanced coin, is always one out of two (a 50 percent chance), no matter what the results of previous flips have been. Each event is independent.
2. and 3. Both of these problems appeared on the National Assessment of Educational Progress second mathematics assessment. In the first problem, only 24 percent of students (age 13) responded with an estimate of 2 (28 percent estimated 19; 27 percent estimated 21). A higher percentage of students was able to compute similar problems; apparently, ability to use a memorized algorithm did not indicate understanding of the quantities involved. The second problem, too, showed a discrepancy between ability to compute and ability to apply computation appropriately. Under a third of the 13-year-olds gave the correct answer, 32; 38 percent answered 16 feet; 21 percent answered 60 feet.
4. This graph (the figures are invented) appears to show a rather large increase in employment between 1977 and 1979. However, the increase

is exaggerated because the graph has been truncated. If the entire vertical scale is shown, the graph looks like Fig. 4.24, giving a very different impression to the reader.

5. There are several ways to solve this problem. A straight computational approach would suggest taking 15 percent of $6.28, adding that amount to $6.28, then subtracting the total amount from $10.00. However in the actual situation described here, an individual is more likely to use the following process: 10 percent of $6.28 is about 60¢ and 5 percent of $6.28 is half of that or 30¢. So the tip is 90¢; add that to $6.28; that is close to $7.00, so your change from a ten dollar bill is about $3.00. If your purpose is to make sure you will have enough money for train fare home, this calculation is adequate, efficient, and shows a clear understanding of the numerical relationships involved.

6. A statement similar to this one appeared recently in a large urban newspaper. A runner reading this article immediately recognizes that, while an 8-minute mile is a reasonable pace for a good runner, 1.6 kilometers in a minute well exceeds the world record for that distance!

7. Some people can immediately visualize the solid object that would

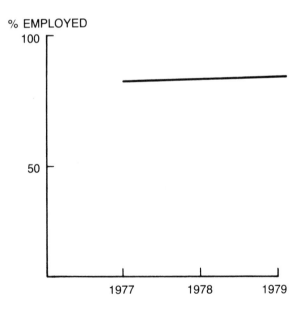

Fig. 4.24

result here (a cone). Others may reason more slowly about what shape must result.

8. Any sequence of numbers can be continued in a variety of ways. While the common response for the sequence 2-4-6 is 8, the sequence might actually be, for instance, 2-4-6-9-12-16-20 (add 2, add 2, add 3, add 3, etc.) or 2-4-6-4-2 (add 2, add 2, subtract 2, subtract 2, etc.). The second sequence might be continued 1-2-4-8-16-32 (double the previous number) or 1-2-4-7-11-16 (add 1, add 2, add 3, add 4, etc.), for example. While there are undoubtedly numerical sequences that can be made from the third sequence, the numbers are also in alphabetical order; using this rule, the next number in the sequence could be 6.

Motivation and Autonomy 5

Behind the phrases "motivating children to learn" or "making mathematics interesting" are the clear assumptions that children, by nature, avoid learning, and that mathematics, by nature, is dull. Special interventions are seen to be required in each case: some kind of pep talk to convince a student to do a particular task, or perhaps some sweetener to make an activity more attractive. Because we value the outcome but expect the child not to gravitate naturally to the process, we devote specific attention to convincing the child to go along with our plans so that later.... So goes the story.

INTERNAL MOTIVATION

In this book, motivation has a very different meaning. It has to do with a person's view of his or her role in the world. "Am I an active participant, or do things just happen to me?" Feeling active is not something we can talk a child into; the feeling must come from real power. Must a child always feel a sense of power to engage in an activity? What about the inactive role people take in front of the TV? Clearly, one need not always be active or controlling in order to learn, or even to enjoy oneself. Still, choosing the observer role from time to time is quite different from being assigned that role as a matter of course.

While giving a workshop on computers in the special classroom, one of the authors of this book began by demonstrating activities with computer graphics. A teacher attending the workshop asked him to go directly to reading or mathematics. The children in her classrrom, she argued, were so far behind in these basic areas that she had little time for art and music. They were important as breaks from the routine, as "motivational devices," but were not the lesson content itself. This sense that the purpose of motivational devices is for managing children's behavior works directly against the notion of children developing autonomy and managing their own behavior. The teacher's request carried a further implication—art and music are enjoyable, but the basics must come first. This raises two other crucial issues: 1) What role does pleasure play in the educative process? and 2) How does one judge what is basic for a particular student in this new age of computers?

From a scientific point of view, the notion that children don't learn when they are bored or unhappy is just silly. They certainly can learn, and the experiences of generations of children testify to this. On the other hand, it would be equally silly to claim that enjoyment of learning doesn't matter. People resist situations in which they are bored or unhappy. The more resistance we face from our students, the more effort we must put into getting the students' attention and cooperation.

The issue of resistance speaks mainly to our own jobs as educators. Life improves for us if our students like what they are doing. And then there are the psychological risks to our students in the use of force. The more we force people to do things that they would choose not to do, the more we are teaching them to be pliant and passive. Can we never, then, ask a student to do that which the student does not choose naturally to do? This, too, would be a silly stance, and has led in some circles to the uncritical acceptance of mediocrity. Excellence requires hard work and perseverance.

If motivation is external (force or sweeteners), we risk teaching passivity; but, without challenge, our students will not achieve excellence. This leaves us with the task of preserving and fostering internal direction. In the fortunate majority of circumstances, young children, especially before they enter school, have exactly the kind of direction that we would like to see preserved in older children and adults. They try "hard" activities and drill themselves relentlessly until they can accomplish them. Of course, this rarely looks like work, but it isn't always pure fun and games either.

Bumps and scrapes, even the rough falls off a bicycle, do not deter the drive to mastery.

Promoting internal motivation involves promoting a sense of self-control, inner power, autonomy. A person may be willing to struggle with a task that does not come naturally and to learn from that struggle. However, if a major portion of the tasks that a person is asked to perform does not come naturally, or is of no felt importance, the person either resists or, in complying, becomes passive. To grow actively, a person requires access to an environment in which he or she can successfully engage in a fair diversity of important, interesting, exciting, and enjoyable activities. Each success fosters a feeling of power and can serve as the springboard for work at more complex and less immediately rewarding tasks. The alternative leaves little energy to risk and explore.

Computers can provide access to a wealth of activities that are otherwise closed off to children—especially handicapped children. A child can do mathematics beyond acquiring some portion of it as a go-nowhere skill. Tammy could draw; before, she could not. Eric could talk fluently enough to make vocabulary acquisition genuinely useful to him. Because Eric's future does not seem closed off, and because Eric's present options are very wide, his natural motivation to explore keeps him striving for new achievements.

By using the computer as an aid, so much more becomes accessible to the handicapped child—music, art, written stories, explorations in science—that the task of the special educator need not focus so heavily on getting the student to do exercises which, though intended to be beneficial, are expected to be difficult and unappealing. Computers can, perhaps for the first time, let us overcome the problem of passivity and change the role of educators into a job of preparing people for full and active participation in social and professional life.

LEARNED PASSIVITY

Autonomy tends to be discouraged in handicapped individuals. Children whose activities are unusually limited, who are watched over and protected, who for years cannot do what they see their peers doing, who are treated as patients, invalids, babies, incompetents, or bystanders learn to be passive. When the handicapped child gets older, this passivity is often enhanced by depression, upon realizing that he or she will never be like his or her parents. Motivation is the basic educational issue for these children.

Handicaps that are specifically school-related—mild learning disabilities that are invisible and largely non-interfering outside of the context of school, for example—may have a different effect on the child's self-image. Except in school, the child's choices of activities are not badly limited, and may be largely indistinguishable from those of other children. The future appears open. In this case, it is specifically school that is the enemy. Though the dynamics are different, the problem faced by the child's teachers may still be largely one of motivation. School is a site of failure and the child's defense may be the avoidance of information and interaction that Margaret Riel studied.

Because of the prevailing passivity, the aggressive behaviors of children with disabilities often come in the form of resistance or obstruction. Even when this is not the child's intention, the behaviors are often either treated as, or converted to, a kind of antisocial fighting back, and are systematically stamped out, if possible. The logic is clear—the antisocial behavior is not tolerable and is not in the child's best interest, ultimately. Unfortunately, the message is also clear that even greater passivity and compliance are desired.

One of the reasons for developing a passive lifestyle is described by June van Lint, in her book, *My New Life* (see Publications in the Resources Section). Mrs. van Lint pictured herself as having been "a normal housewife, well occupied with home, husband, four adopted children (ages 2 to 9), PTA, choir, and the church couples' group." Then, as a result of an accident in their station wagon and ensuing complications, she was left almost totally paralyzed, unable to control voluntary movement from the neck down and also unable to speak. Especially during her hospitalization, but also at home, she was taught to be passive in numerous ways. One of the most powerful lessons resulted from having her attempts at communication misunderstood or ignored. She learned, for example, to accept pain when people were mishandling her rather than to try to inform them of their errors. Because she was unable to communicate the location of the pain, any show of pain would cause the others to slow down, trying, usually unsuccessfully, to figure out what was bothering her, with the result that the painful procedure was prolonged rather than relieved.

Passivity made Mrs. van Lint's life easier, but at a cost. Her family's support of her efforts at self-direction and her memory of earlier autonomy gradually helped her move from the role of passively minimizing pain to regaining active involvement. Mrs. van Lint's lessons in passivity were sudden and harsh, but she was also an adult who had previously learned autonomy. Unlike the child who is handicapped from a young age, Mrs. van Lint knew what she was missing.

To deal with a complex world, the handicapped child must be less passive, not more so. Parents, teachers, and others cannot possibly anticipate all the problems that will be faced by the disabled individual—much less explicitly teach the necessary creative responses. Thus the disabled child must not only be given the academic and intellectual skills to respond adequately, but he or she must be prepared emotionally as well. The more active the school, social, and working life of an individual, the more complex will that person's life problems be, and the more the solutions to those problems will depend directly on the autonomy of the individual. A less pervasive disability might be expected to carry less of a penalty in diminished autonomy with it, but the issues remain much the same.

All well and good, you say, but if Johnny can't read or Jane can't make change, how far can they get, anyway? Besides the world isn't all fun and games; self-discipline is also an important emotional strength. True, and for this reason it may be necessary to make self-governance a curriculum in itself: autonomy as a "basic." In contrast, if we begin instruction with mathematics or reading before they are of felt value to the child, a part of the hidden curriculum is learning-to-do-what-you-did-not-choose-to-do. However, if instead, we begin by creating an environment that attacks a child's passivity, we have the best chance of maximizing the child's future both in school and in the adult world. Some children, at least initially, are so disconnected that any activity that engages their attention and effortful participation is valuable almost regardless of its content. But in most cases, it is possible to introduce activities that are rich enough to provide standard curricular content as well.

By providing increased access, the computer lets a child exercise choice and power—assertive behavior that is social, not antisocial. The computer opens the door, but how do we help an already passive child to develop an "aggressive" choice-making stance and spend the effort to go through this door. For some children accustomed to a passive role, the new environment must be made especially attractive to encourage learning the new active way. One route is through music and art.

COMPUTERS, MUSIC, AND ART

Most children love art. They like to represent the images they have in their heads by putting crayon or pencil to paper. They enjoy bright colors and rhythmic designs almost regardless of the medium they are using— whether they are painting or creating a mosaic of colored cloth. Children also generally love music. The toddler dancing while a record plays, the 3-year-old quietly singing to herself, the boisterous playing of the nursery school Orff instruments, and the teenager's devotion to the radio all attest to music's compelling force. Art and music are exciting to students both in elementary and secondary school. They are also rich in personal, emotional, and intellectual content, and therefore may be an important element of educational plans even for children who could succeed with activities more intensive in language or mathematics. Further, they are easily relatable to other physical and mental activities. As a result, they may represent the ideal highly motivating environments within which to begin an educational program.

Such are the reasons for encouraging art and music activities, but why particularly with the computer? First, the computer can enhance performance. Second, for those handicapped children who cannot use the traditional materials, the computer can provide access. Third, by enriching what is possible, the computer adds opportunity for experimentation and intellectual growth that can lead both directly and indirectly to other activities.

Geoff and Laurie are officially registered at the Peet School for Deaf Children, but they now attend most of their classes at the public junior high down the road from Peet. In the year before they joined the mainstream program, they spent a lot of time working with Peet's computer, much of that time using the computer's mail program to send letters to each other and to other students and teachers at Peet. Their teachers attribute much of the academic growth they showed during that year to their involvement with the computer.

For both Geoff and Laurie, their introduction to the computer did not stress English, but rather allowed them to draw. Both were extraordinarily quick to pick up the computer commands (Logo) that moved the turtle on the screen, creating their beautiful graphics. Geoff worked laboriously on what might have been a Swiss chalet, an intricately designed house, complete with scalloped shutters, curtained windows, and trim around the sides. He explained that his family spends ski vacations in the winter in a house like that. Laurie tended more toward geometric patterns, but also worked for a long time on a remarkably life-like tree. Some of her geometric patterns reminded her of snowflakes, so she got the idea to make Christmas cards with her designs using the graphics printer.

For both of these children, graphics was a natural introduction to the computer. It had appeal. It allowed them to be creative. The computer represented a new and complicated technology which, if mastered, could be a source of legitimate pride. Art allowed them to approach this new technology from a territory in which they had no handicap. It established the computer as a tool with which they could be successful and could show their competence to others. Formal lessons offered an opportunity

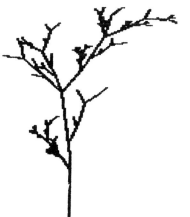
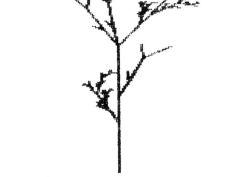

Fig. 5.1 Two of Laurie's trees.

to develop competence and master a territory, too, but, because they applied the expectations, standards, and schedules of others, they demanded greater personal risk.

Enhancing Performance

Artistic creativity is stifled by the "I-can't-even-draw-a-straight-line" mentality. One need not be able to draw a straight line to be an artist. But creativity is also stifled when the artist cannot achieve a desired effect. Sometimes what one wants is a straight line. Free use of color and motion in a painting may not require a computer, but experimentation with rhythms generated by a careful repetition of particular patterns may be difficult for an unskilled freehand artist. Having the computer as one of the available media enables the artist to choose the medium to suit the work. A similar case can be made for music.

Using a language such as Logo, a child can quickly make pictures or graphic designs and can compose appealing tunes. The quickness is also an enhancement of performance. Without waiting for the motoric skills to develop, they can apply their cognitive skills and try out their ideas. For the less academically able child, the quickness—getting the most dramatic effect for the least programming—may be particularly important. Children develop competence with the computer especially fast with graphics, but the least academically advanced child—one who is very young or severely handicapped—may be best started with the robot turtle. This very physical "animate" object is the concrete precursor to the screen "turtle" that is used in graphics activities.

Access for the Handicapped

With the computer, children who cannot write a single letter may be able to do intricate, beautiful artwork, design mechanical devices, draw maps, choreograph a dance, or compose a symphony. Many of the children to whom you were introduced earlier in this book were doing music or art using the computer. Seven-year-old Eric played two short phrases of Tchaikovsky's violin concerto and a 13-year-old girl at Ellis Middle School was learning to compose music that resembled Hungarian folk songs. Tammy drew, and the computer filtered out her unintended movements to allow her to produce the picture she wanted. Such personally chosen computer environments can foster considerable intellectual growth.

For a student whose major problem is distractability, there are more subtle access issues. That student is deprived of certain information while other events interfere with attention. Here, the computer's ability to enhance performance may serve to focus attention better. By speeding up the interaction, it, in effect, leaves less room for the distractions.

Opportunity for Experimentation

The self-imposed "curriculum" of the pre-school child includes not only tremendous amounts of practice, but also a considerable amount of remarkably systematic experimentation—a kind of research in its child-like form. By contrast, students in school tend to think that research means looking up what somebody else (probably even the teacher) already knows, and copying—sometimes without even understanding. At the other end of the spectrum from slavish copying of an experiment is a kind of exploration that is unrestricted, but not sensitively informative either—a kind of trying-things-out that is neither systematic nor rule-bound. The youngster with a chemistry set is often trapped by one of these two extremes: either follow the book and rediscover the predicted, or mix things on your own and, for the most part, discover nothing.

Somewhere between these lies true scientific research, as an example of serious academic pursuit and intellectual stretching. Students labelled "special" are no less in need of such experiences than are other students. Their typical situation may even leave them more in need of the cross between constraint and open-endedness that constitutes a challenge.

Art is not generally thought of as a "research" activity, but it has the virtues of open-endedness and is, for many students, highly attractive. For the true artist and musician there are genuine rules and constraints—sometimes self-imposed—guiding a composition. But beginners have so much difficulty handling the medium—getting a sound out of the flute or painting a picture—that sensitive experimentation is far beyond their abilities. The computer's ability to reduce the time and technical skill required to execute a graphic or musical composition makes it possible even for beginners to perform real experiments in these inherently attractive areas, imposing self-created rules and constraints, and using the results to guide new explorations. The Ellis school student who was trying to compose a Hungarian-like folk song was doing exactly that kind of research.

So is Alejandro, a 17-year-old exchange student from Spain. His work with the computer began with art. This moved him into complex mathematics, including some concepts from calculus and, as a by-product, into some important social interaction.

Alejandro is bright and has been an exceptional student throughout his schooling. Although he is doing well in the progressive suburban high school to which he came, his difficulties with English make his success come at the price of very diligent work and long hours. The time he must devote to study as well as his lack of ease with English both limit Alex's social interaction. But Alex loves graphic design and has used his Introduction to Computers course to research a family of curves he happened on by chance while trying to generate a spiral in Logo. He went on to analyze the mathematics of the shapes and began to predict the patterns they would make. His shapes attracted the attention of several other students in class, and have drawn them to him. Though their talk focussed initially on his work, it became a bridge for further social contact.

COMPUTERS AND MOTIVATION

Although the route to academic learning through activities like art and music may seem somewhat indirect, it tends to be a very accessible route. The computations a computer does in the midst of solving a mathematics problem are invisible, but the turtle tracks left on the screen provide a record that one may simply appreciate, or may retrace and analyze. Posing one's own problems as Alex did, is, of course, an indicator of an active intellect. But each opportunity to do such problem posing also further strengthens an active rather than a passive intellectual life style. As in Alex's case, problem posing in what began as a purely esthetic endeavor can provide sophisticated academic lessons.

Within the scope of this chapter, it is not possible to give prescriptions stating how to help every child overcome intellectual or emotional passivity. However, the following descriptions and stories of handicapped children

Computers and Motivation

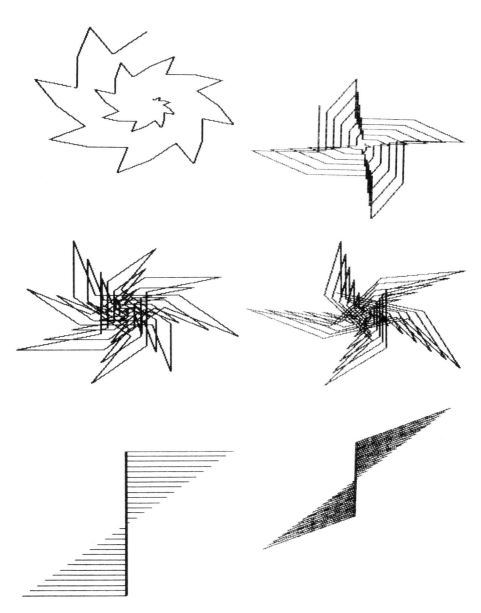

Fig. 5.2 Six of Alejandro's designs.

have been selected to illustrate the wide variety of options computers offer in this endeavor. Included is also a general discussion of the principles used in designing the devices and activities described. These examples illustrate ways that teachers can tailor programs to the needs of particular individuals with whom they work.

There are at least four aspects of autonomy with which the computer can deal effectively: the fostering of engagement, the broadening of horizons, making competence attractive, and the building of self-esteem and a sense of independence.

ENGAGING THE UNMOTIVATED

Sylvia Weir and Ricky Emanuel provided an autistic boy whom they call David with a computer-controllable robot turtle. The robot turtle could perform a few simple behaviors such as moving forward in the direction it was facing, rotating around its middle to face in a new direction, turning its headlight on or off, and beeping. David learned to control it remotely by pressing buttons that were labeled with pictures of arrows for the different kinds of movement, a picture of a bugle to symbolize the horn, and pictures of a headlight shining or darkened.

Fig. 5.3 Harvard Associates' Tasman Turtle.

Each press of the forward arrow moved the turtle a fixed distance, about an inch, forward. Each press of the right-turn arrow rotated the turtle fifteen degrees to the right. It was also possible to multiply the effect of a command by preceding it with a number, for which there were buttons as well. Finally, it was possible to build a complex behavior out of a sequence of simple behaviors, and to assign that complex behavior to a single button, thus creating a new command, let's say "box." For example, one might "tell" the turtle that this new command consisted of going forward five "steps" (inches), turning right 90 degrees, going forward another five steps, turning right 90 degrees, and so on, tracing out a square until the turtle was back precisely where it started.

During their seven sessions with David, the experimenters noted that when he was controlling the toy and determining his own agenda for the session, he showed obvious pleasure, appropriate and spontaneous speech, improved vocal tone, and a bodily posture indicative of interest and involvement. When he was not in control—for example, when he was being taught—he assumed what they called a "passive pupil role" and "acted autistic."

One of the authors of this book (Paul Goldenberg) provided a similar setting for Nancy, a 5-year-old psychotic girl. Nancy reacted very much as David did to being in a situation she could fully control. At the Logo Laboratory at MIT, she got a chance to play with a robot turtle similar to the one used by David, but did not have a suitable button box available. Instead, the computer terminal was used as a source of switches for controlling the turtle. Each turtle action was represented by a single keypress (e.g., F for "forward," R for "right turn," T for "toot," etc.).

Although Nancy knew her letters well, she appeared to have difficulty attending to the behavior of the turtle and also keeping track of which key to press to cause an effect, with the result that she played almost as much with the many functionless keys, especially M, as with the roughly half dozen that caused the turtle to react. To help her, Dr. Goldenberg stuck white paper stickers on the few important keys. On each sticker, he printed in red the letter corresponding to the key it covered. No new information about the meaning of these keys was conveyed by the labels, but their location was effectively highlighted. The added visibility helped Nancy attend to the functional keys among all the others.

Nancy still played with other buttons, but her moves seemed quite different. After a few presses of the labelled F key to make the turtle

Fig. 5.4 Keyboard with letters F, B, R, L, and T highlighted with self-sticking labels.

move forward, she deliberately switched to the unlabeled (functionless) M key and pressed it once or twice, looking closely at the turtle to see what would happen. She switched back and forth between labelled keys and the M key, always watching the turtle. After several trials, she made a label for the M button just like the other labels. She stuck on the new label, as Dr. Goldenberg had done, and returned to playing with the various keys. Although one onlooker described the label-making as off-task imitative behavior, it seemed like a perfectly reasonable experiment to have performed. She was attempting to make M work. Apparently, her experiment provided her with information that verbal explanations had failed to convey—she stopped playing with the unlabeled buttons and focussed on the buttons she knew had an effect.

She learned that she could control the turtle's behavior by pressing the right buttons. She commanded the turtle to come to her, to run away from her, and to knock over a tower of small wooden blocks. As Nancy made the turtle act out her wishes, Dr. Goldenberg narrated the events, "humanizing" the turtle by characterizing its actions as "talking to her" when it tooted or "running away from her" and "coming to her" as it moved back and forth. Nancy's reactions both to the toy and to the narration were intensely positive. When she repeated a command, she gleefully applied Dr. Goldenberg's description of the event, and sometimes embellished it with her own. Frequently she said "I'm so happy at that turtle; that turtle listens to me," or "Aren't you happy at me!"

From a child who entered the laboratory holding onto her teacher tightly with her left hand and pinching her eyes shut with her right, emerged a child who was clearly in control, at least for the moment, both of the turtle and of herself, a child who could communicate ideas to this toy and could see consistent and predictable responses to her communicative efforts.

What is it about working with the turtle that creates a rich and useful experience? The immediate responsiveness of this mechanical toy makes its behavior readily transparent even to David and Nancy, for whom communication is not an orderly or easy process. Playing with the turtle initially involved little or no demand on David and Nancy other than pushing buttons, a fact that was considered significant for their acceptance of the toy. But understanding and being able to control the behavior of this responsive toy gave them an experience of power, of being in charge, that they rarely had in other situations.

Playing with the turtle generalizes well to other educational areas. There is mathematics both in the planning of its moves and in the language by which it is controlled. Perhaps more immediately important for Nancy and David is the communication involved. Controlling the turtle in a purposeful way implies knowing the correspondence between particular buttons and the behaviors they symbolize, recognizing the means-end relation of buttons and turtle behaviors, being able to predict the outcome of a sequence of button-pushes before executing them, and having a purpose to communicate to the turtle. Although this button language for the turtle is vastly simpler than speaking English to a person, David and Nancy employed four skills basic to communication and often not observed in the autistic child's behavior with people:

- the knowledge of shared symbols
- application of symbols as a means to a predicted, purposive end
- sequencing of symbols to generate new effects, and
- being the initiator rather than (only) the responder

Perhaps it is precisely because the button language is so simple that the child can exercise these fundamental skills and then later learn to assimilate the harder skills of communication with people. David and Nancy appeared to be taking those steps while enhancing rather than diminishing their sense of autonomy.

For David and Nancy, an important feature of the computer was that it provided them with an obedient servant. These children were more or less free to initiate explorations in an easy-to-control environment, free from a history of failure, and free to learn how the computer would respond—important features to look for when providing such an environment for any child. This freedom capitalizes on a kind of play behavior that children engage in even at very early ages. For these language-impaired children, the cost of communication is high, and so developing the motivation to communicate is a first priority.

As a technique, this may be important in establishing the utility (or even possibility) of communication for the child. If the child does not perceive the environment as being responsive to his or her wishes and needs, that child will have little reason to try to encode those wishes and needs in a consistent way to inform the environment. Just as the preverbal babbling of the infant gains social "meaning" to the extent that it elicits a social response from the adult, growth of communication in infants and very young children may depend more highly on the reactivity and responsiveness of adults (and other children) than on the examples set by them. A language pathologist, teacher, or parent can certainly employ responsiveness as a primary teaching technique, but there is something to be said for allowing a machine to take a major part of that role. With both David and Nancy, the simplicity of communication with the machine was first extended by imitating that communication with the self or the observer/therapist. Both children then augmented it by making spontaneous comments to the therapist, mostly to call attention to the event. Thus, although the communciations began with the machine, subsequent communications with people were not mechanized, but served a truly social function.

Children are most spontaneously communicative to adults during an acitvity in which they and the adults are both engaged. If the activity is designed properly, what a child does with the computer is, in itself, expressive of the child's own intentions, and is personally controlled and active in a unique way. It practically compels communication: children ask how to tell the turtle to draw in a different color or to repeat a particular line, they tell their friends how, they call to others to show them a pleasing effect, express delight, comment on how dumb the turtle is when something unexpected happens, and so on.

WIDENING HORIZONS

Unlike David and Nancy, Jay, a severely motorically disabled cerebral palsied adolescent, had no problems with understanding cause and effect. His problem was extremely restricted mobility. By functioning as a prosthesis, the computer made possible activities that had previously been impossible for him.

Jay had been taught to type laboriously by pecking out each letter with a wand strapped to his forehead. Most of his practice was gained in school exercises, but he had occasionally used the skill to write short letters home, at the initiative of the school. During Jay's first session with the computer, he learned that by typing particular instructions at the keyboard he could command the computer to draw pictures that he designed. Though his typing skills were very poor, he reacted with tremendous enthusiasm and resisted even being taken away for lunch. A week later, when he returned for a second session with the computer, his typing had improved dramatically. The computer did not teach him to type—during the intervening week, Jay had lived ninety miles away in his residential school and had no access to the machine. We do not know how much of Jay's typing improvement was due to the practice he gained writing a long letter to his parents about his computer experience, and how much to the new discovery that typing gives access to a highly motivating activity, computer programming. What we do know is that something about Jay's motivation changed as it had never done before.

The most far-reaching importance of an experience such as Jay's is neither his designs, per se, nor his improved typing skill, but rather the unusual length of the letter to his parents. There is, in general, not enough about his institutionalized life that is worth the extreme effort it takes to type it out—the major issues are, for the most part, already known by his parents, and the daily events change very little. His generally poor performance in communication and its component skills was not, as appeared to be the case with Nancy and David, a cognitive problem in manipulating the symbols. However, as with Nancy and David, the first educational issue to be dealt with is again not building new skills. For Jay it appeared that the stimulus to communicate grew directly out of having broadened his experiences. The bargain that had previously been made with Jay—"If you gain new skills (first) your world will (then) get richer"—is rarely convincing. On the other hand, Jay's encounter with the computer was more like: "Here's a new experience: deal with it!" That approach often gets people moving.

MAKING COMPETENCE MORE ATTRACTIVE THAN INCOMPETENCE

Learned Passivity

The Laura Bridgman Center enrolled forty preschoolers, a half-and-half mix of able-bodied and physically disabled children of normal intelligence, vision, and hearing. Many of the disabled children required a crutch or a brace and some physical therapy, but were otherwise able to walk and had full use of their hands and mouth. Four of them, Anneke for one, were not so physically able. Anneke entered preschool when she was 5, exceptionally small for her age, unable to walk or to talk (or chew and swallow effectively) and unable to make much use of her hands. She signalled Yes by raising her left arm and No by raising her right, and the other children, young as they were, learned to understand her left-right code. She impressed her teacher Lise as being quite bright because she listened to and apparently understood everything and, even without speech, exhibited an outrageous sense of humor. Anneke was a determined 5-year-old, socially conscious of the other children and eager to do everything that they did.

One day, Anneke was seated on the floor, propped against the wall of the hallway, and Lise was squatted next to her, feeding her a snack. As Lise left to refill Anneke's drink, she waved an admonishing finger at Anneke who had not been seen even crawling and, with a teasing smile, said, "Now, don't you go anywhere while I'm gone!" When Lise arrived back with the drink, the hall was empty. When Lise called Anneke's name in surprise, she heard a chuckle and a scuffling noise from around the corner. Anneke had pushed herself up and, leaning her back against the wall, had sidled along it until she was out of sight.

Anneke flourished in the Center, and grew particularly attached to Lise. She learned to point to any letter Lise named, to spell her own first and last name, and to read several words. Her oral control remained totally inadequate for speech (and, little doubt, always would), and her manual dexterity was, at least at present, inadequate for signing or writing, but she was learning a more complex system of pointing to augment her communication.

Two years later the picture was not as bright. Not only had she grown too old for the Bridgman program and thereby lost her favorite teacher and that exceptional environment, but she was enrolled in a special school in which most of the other children had multiple handicaps. Just as she had tried to "fit in" before by imitating the other non-handicapped children, she was now learning how to be a retarded, passive, incapable child. She again had an excellent teacher, but expectations—both the school's perception of Anneke, and her own perception of social demand—were different. At age 8, she shows no signs of reading, recognizes letters erratically, cannot reliably spell her name, and has the school doubting her potential.

Seven or eight is about the time when many severely physically disabled children begin to see their inabilities as inescapable. Able-bodied 4-year-olds do not consider it abnormal that they cannot do many things that adults can do. Neither do disabled 4-year-olds, for the most part. By seven or eight, these same children may become depressed as they begin to see "no hope." This realization must be anticipated, and its impact mitigated if the child is to develop to his or her fullest. Contact with similarly disabled but functioning adults helps. If one cannot grow up to be exactly like mommy or daddy, at least there is someone else to be like. Also, experiences of capability and control help. This is where the computer comes in.

Capability, Control, and Mastery

Tammy (you met her before in Chapter 1) had often had crayons or magic markers strapped to her hands so that she could join others in her class in "creating art." Perhaps it had been assumed that she would find the activity of coloring satisfying, regardless of the outcome—Tammy had long been thought to be retarded—but the result had always been scribbles and frustration. When the computerized drawing tablet was first presented, the associations were too strong, and Tammy was very reluctant to have anything to do with it.

By contrast, Tammy was positively excited at the possibility of playing TV ping-pong, a presumably harder task, but one in which she had no history of failure. By setting several parameters, it was possible to vary the difficulty of the ping-pong program to make it challenging yet playable. A filtering technique allowed Tammy to control the game despite her athetoid movements.

At first, Tammy missed a lot of the balls, but she persisted despite her initial failures. When asked if she wanted the game to be made easier, she replied that she did not. It did not take her long to develop some new techniques for manipulating the joystick that controlled the line representing her ping-pong paddle, and within about ten minutes of play, she was being quite reliable. There was, however, something remarkable about her strategy. After she became skilled at controlling the paddle, she tried to anticipate the position the ball would be in when it arrived on her side of the screen. However, for almost a half hour, she seemed

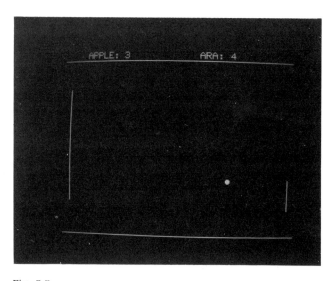

Fig. 5.5

not to recognize when the ball would hit the ceiling or floor of the screen and thus bounce and change its direction. It was as if she did not understand bouncing. But of course! When had she ever bounced anything? Yet, before an hour was up, she began routinely beating the machine at ping-pong.

And she loved it!

Finding an activity that Tammy loved allowed for a view of her when she was putting out effort. Her intelligence had not shown through in the past, perhaps precisely because there was little to apply it to, and little reason for her to waste her energy. In the ping-pong game, she could be seen adjusting her strategies quickly. Her cleverness, understanding, planning, and adaptability became evident, as did what knowledge she did and did not have.

The physical skills that Tammy developed—the ability to manipulate a spot of light on a TV screen quickly and accurately—could be applied to a variety of other tasks, but the most important first step had been taken. There was now an activity that she enjoyed and really worked hard at. If enough of those could be found, Tammy would then become a chooser rather than a passive recipient of services, a poor beggar.

The computer was instrumental in providing Tammy with a chance to perform the sort of informal physics experiments—investigating, the nature of bouncing—that are done by able-bodied children during their outdoor activities. The computer can also provide a medium for learning formal physics. Tammy's computer environment can be continually redesigned as her skills and interest progress, becoming, if desired, a highly sophisticated, "well-equipped" physics laboratory.

As with the ping-pong game for Tammy, computer games may be selected to exercise other especially useful skills. A game in which one explores a dungeon full of treasures and monsters, for example, can provide a great context for reading short bits of text, and might be an excellent motivator for children who need extra practice in reading. People resist doing tasks that appear to have no value whatsoever. If something does not appear to be useful, then it must, to be valued, at least be pleasant. The problem for many severely disabled children is that not much looks very useful. It is within our technological power to make it possible for Tammy and others like her to hold a variety of jobs, but unless their education changes radically, this possibility is not likely to bear fruit. From Tammy's chair, nothing seems accessible, and therefore there are no skills that appear useful to her. From Tammy's point of view, play is at least fun. From our point of view, it may also be her key to a new life.

Tammy had gone through a stage like Anneke's and lived as a "retarded" child for six or seven years before she began to wake up again. Tammy's initial motivation came from computer games like her ping-pong, but computer aided freehand drawing held a special fascination for her, as it was clear proof that there were activities she could do that had before seemed totally impossible—an important lesson.

Self-Discipline

At times, and with considerable satisfaction, David, the autistic student with whom Sylvia Weir and Ricky Emmanuel worked, gave himself commands by pushing his belly button, saying the instructions out loud, and then executing them just as the turtle had done. One interpretation of David's behavior is that he needed the verbal activity to gain control over his own behavior. Donald Meichenbaum (see Publications in the Resources Section) explicitly uses such verbal self-instruction as a technique for helping children gain self-control. During ongoing play activities, he teaches impulsive children to talk to themselves. Dr. Meichenbaum gives the example:

> While playing with one hyperactive child, the therapist said, "I have to land my airplane, now slowly, carefully, into the hanger." The therapist then encouraged the child to have the control tower tell the pilot to go slowly, etc.

The technique makes a deliberate attempt to help the child build up a verbal repertoire of self-instructions. Activities are chosen in which the child can feel competent, and which are not associated with failure and frustration.

Although Dr. Meichenbaum was not using computers, the parallel is compelling. Dr. Meichenbaum's cognitively based therapy adds language to behavior, using "the child's conscious self-regulatory ability." The focus of Dr. Meichenbaum's intervention is to apply the technique of thinking through first and then moving step-by-step to self-regulation. By teaching the child a strategy based on self-governance through the use of language rather than a conditioned response to externally imposed behavior modification, the child is learning autonomous control and a model for behaving intelligently.

Building Self-Esteem and Independence

Defending the computer department's budget at a parent/school-committee meeting, Gregory, a high school junior announced that the computer department served more than the students who were currently taking computer courses. He, for example, types all his English papers on the computer and uses its word processing capabilities and spelling checker to help him improve his writing. "And now," he continued, "I get As on my English papers, which is quite a change, since I spent many years going to special schools because I am dyslexic." Perhaps the most powerful message delivered by his announcement came from the aftershock. Three of his teachers who were present had not even known that he was dyslexic, his performance was so totally normal.

Gregory had largely circumvented the problems caused by his dyslexia and was even becoming proud enough of his accomplishments to name his disorder in public. But for many years, he had been labelled according to his disability, and not according to the many outstanding abilities he had.

The 13-year-old composer at Ellis middle school has been more fortunate. He, too, is dyslexic, and he is, as all of us always are, trying to strengthen his weak spots while capitalizing on his strong points. But his psychological position is different. He has, from the beginning, been succeeding at the vast bulk of his work. Like Gregory, he is expectionally bright and talented. Unlike Greogry, it is his talents, and not his dyslexia that has been his defining characteristic. Still, he is sometimes frustrated at needing so often to have a study partner, someone to read to him. Even in mathematics, at which he generally excels without much reading, there are numerous word problems, which are an irritating reminder of his dependency. Musical composition is different. Though he has some visual sequencing problems that infrequently get in the way of his sight reading, he has developed strategies that allow him to be largely independent. The computer has played many different roles in his development. It can let him experiment with music he could not perform; he especially values the ability to edit his musical compositions, but most of all, the computer can transcribe his finished compositions for him, which gives him another source of independence.

Talent in writing, mathematics, music, and art are all respected highly enough to be suitable reminders to an otherwise handicapped child that he or she is not a total failure. All four talents potentially draw from one's intelligence, creativity, and problem-solving ability. Music and art, however, do not rely in the same way on arbitrary visual codes. Also, they are typically not saddled with the academic image that often poisons other endeavors for children who have already failed a lot in schools. Hence, music and art may become areas in which they can easily shine.

Of course, music is not the only way to build up self-esteem, and computer graphics is not the only way to expand a child's world. The pairing between activity and child in the preceding examples is only suggestive, not prescriptive. When choosing an activity to meet a child's needs, consider the availability of equipment, the physical abilities of the child, the child's interests and inclinations as well as they are known, and by what tool the particular area of motivation most needful of help may best be served: the servant turtle, a more flexible world-expanding tool, etc.

CONCLUSIONS

Any activity that varies the environment has potential for stimulating learning and enhancing motivation. Those that provide the experience of control are particularly important. The computer is uniquely suited to providing playful and non-playful experiences of being in control. Even for the normal child, making something interesting happen, or having control over an interesting event, is absolutely thrilling. A computer's flexibility allows it to be tailored according to the child's tastes and abilities. It can respond with whatever behavior is most attractive to the child (e.g., speech, music, graphics, movement) and to whatever signals the child can use (e.g., speech, typing, movement, proximity).

The computer is highly useful in teaching reading, writing, and arithmetic. It is a promising new tool in assessment. But it is through its ability to motivate activity and interest that it may have its greatest potential for special education. It is here that, for the first time, we may be able to conquer the problem of social, intellectual, or academic passivity and change the role of special educators—even of the severely handicapped—into a job of preparing people for full and active participation in business, law, teaching, science, architecture, counselling, music, art. . . .

Assessment 6

Recently a 40-year-old student, reentering school after years of working and raising children, found herself taking the Graduate Record Examination. Although she had studied the mathematics that she had long since forgotten from her high school days, having to answer so many questions in so little time overwhelmed her. Despite her preparation, she did poorly on the test. While solving practice problems, she had slowly convinced herself that she was able to think about mathematics sensibly. However, the speed required by the exam itself prevented her from using the thinking skills she had begun to believe she had.

Nancy, described above, is a real person of our acquaintance. She is also a paradigm of many of us who feel that testing procedures often sell us short. Nancy does not have a classifiable special need, although it would not be difficult to list several that affect her ability to perform: length of time since formal school experience, anxiety, perhaps learning style.

Assessment is a delicate matter. Elements of the testing situation that have nothing to do with the content of the assessment often have significant effects on the responses of those being tested. Almost all of us have special needs in the testing situation, including the need for support, the need for self-confidence, the need for time, the need for appropriate tools with which to make clear to the tester what we understand, and the need for a reason to engage in the testing situation sufficient to motivate us to grapple with it.

Educators who are thoughtful about the testing they do are often skeptical about the amount and quality of information that tests provide. Group-administered, standardized achievement tests may be appropriate for comparison of large groups, but many teachers find that they are not helpful for acquiring information about an individual child's strengths, needs, and learning style. The results of such tests are not specific (a percentile score gives little indication of particular strengths or difficulties within a content area); the content of the test may not match the content or priorities of a particular curriculum; scores may not be available for many weeks after testing. The large amount of criticism that has been aimed at standardized testing will not be detailed here, but the damage that scores from such tests can do may be summed up by this angry statement of a sixth grader, learning about the normal curve used for test construction: "You mean 50% of the kids have to be below average on this test? But that's not fair!"

In all educational settings, sensitive teachers have found the need for other measures of student progress: teacher-made tests, standardized individually administered tests, observation, samples of children's work, and self-evaluation by the child. In general, the more special needs the child has—that is, the more he or she differs from the average child for whom the standardized test is constructed—the more careful and sensitive the evaluator and the evaluation tools must be.

There are five facets of the testing situation that may affect the quality and accuracy of information gleaned from it: 1) the appropriateness of the content of the assessment to the curriculum; 2) the amount of probing into the process of thinking allowed by the testing situation; 3) the way in which the directions, questions, or problems are communicated to the student; 4) the method of student response; and 5) the attitude of the student toward the testing situation. A sixth facet, which must be considered if quality assessment is to occur, is not a characteristic of the testing situation itself but of the management of information once it is gathered: how the information is recorded, stored, and accessed.

This chapter explores the potential of the computer as it relates to these six aspects of assessment. The usefulness of the computer for standardized testing is evident: computer scoring of standardized tests is one of the earliest and still most pervasive uses of the computer in education. In this chapter, however, we are more concerned about the needs of parents, teachers, and children for tools that will help in diagnosing and planning for the individual child.

ASSESSMENT AND THE CURRICULUM

In selecting an appropriate assessment tool, three questions must be considered: why is the test needed, what is to be learned from the test, and what form of test is likely to give the greatest amount of information? Testing is used for such diverse purposes as choosing goals for remediation in mathematics, judging creativity in writing, probing ability to correctly wire a circuit board, assessing understanding of the major themes of the industrial revolution, seeing if the spelling of ten particular words has been memorized, and so on. Perhaps the most difficult task in assessment is selecting a tool that matches both the overall curriculum goals and the individual student goals in a particular school or classroom. Teachers often find themselves in the position of being required to give tests that do not tell them anything they do not already know; parents often see a test score reflecting too little of the complexity of their child's thinking.

Special educators require good assessment tools that will aid them in monitoring progress of children who have a large range of special learning needs. Children who have educational handicaps, their teachers, and their parents are continually seeking information about how these special individuals "measure up" to the usual expectations about school accomplishments. Assessment of individuals with special needs should not underestimate either their strengths or their needs. Formal or informal assessment must give clear, useful information that assists in the planning of appropriate curricula, while the characteristics of the assessment itself produce as little anxiety as possible.

Developing a repertoire of flexible assessment tools for the diversity of students and situations within even a single school setting is a demanding task. An elementary school teacher who is responsible for the progress of twenty-five students in all school subjects or a high school science teacher who sees 150 students each day is often forced to make compromises in the amount and quality of ongoing assessment. Ideally, assessment and teaching are inextricably linked; good teachers engage in informal assessment throughout the school day and often store incredible quantities of information in their heads about their students' progress in particular subject areas. Nevertheless, most teachers find the need for more formal periodic assessment to gain information for themselves, as well as for the students and the students' parents. Unfortunately, limited time for such testing and the availability of only a few types of testing instruments has often had a deleterious influence on the curriculum—less self-pacing, less in-dividualization of curriculum, less modification of test design to meet individual needs. The testing movement, which Alfred Binet thought would lead to greater responsiveness of the educational system to the individual, ironically has led to just the reverse.

The choice of a particular test instrument by a state or school system often pressures teachers to teach the content stressed on that test. Knowing that her students will take the New York State Regents Examination may lead a history teacher to cut short interesting discussions about issues that will not be covered by the test. A fourth-grade teacher may move his class through long division at a pace that he knows is inappropriate for many of his students in order to cover all required topics by the end of the year; he may leave out the unit on informal geometry that will not be stressed on the achievement test. The tests thus dictate the content and pacing of the curriculum, rather than vice versa.

The child who has different strengths, different pacing, different ways of understanding directions, different ways of making responses is at a disadvantage. The teacher who teaches different content with a different emphasis, at a different rate, in a different order, or individualizes for different children is also at a disadvantage, no matter how good the reasons are for doing so.

COMPUTER ENHANCED ASSESSMENT

The computer can aid teachers in tying assessment more closely to the curriculum, but not by adding new formal tests to those already available. Rather, it can be a tool to aid in the ongoing informal assessment that is an integral part of the teaching process. The goal here is to develop assessment systems that closely match the curriculum in a particular school or classroom and that are flexible enough to support variations geared to the needs and skills of individual children.

The following are two examples of teacher-made computer assessment systems, modifications and extensions of parts of the teachers' own previous classroom practices for which the computer appeared to be an appropriate tool. Since informal assessment and teaching are often impossible to separate (assessment leads to the next step in curriculum planning, carrying out curriculum leads to the next step in assessment) these teachers created programs that supported this teaching-assessment cycle.

An Experiment in Improving Writing Skills

Bob Bender teaches ninth- and tenth-grade remedial writing. Many of the students in his classes begin the year unable to write a coherent paragraph and lacking the motivation to attempt one. In particular, they lack the skills to improve anything once it is written down. Most of these students are used to receiving essays returned to them with corrections throughout, but even then are often unable or unmotivated to make appropriate revisions. Bob instituted a system several years ago in which 50 percent of a grade is earned automatically by turning in anything at all; students then revise their work and receive additional points for every improvement made. Over time, a point system has evolved, which grants small numbers of points for spelling corrections, punctuation, better paragraphing, and so forth, and larger numbers of points for rewriting a sentence so that it is clearer, livelier, more exciting, more interesting, or sometimes just "better." Lively class discussions revolve around which of two sentences is "better" and why; students begin to articulate their own criteria.

This year Bob has access to a microcomputer and, in conjunction with the typing teacher, has set up a new system. Students who learn to type during the year (either by taking a typing course or by using a computer program that teaches typing) receive a certain number of points toward their final grade. Students who choose to work for these points use their own essays as typing practice. They type their first draft into the computer, print it out, and turn it in to receive their usual 50 percent credit. Then they use a text-editing system on the computer to revise their work.

This system has several advantages. For the students, the most important factor is increased motivation. First, the ease of revising without recopying made possible by the text editor encourages students to attempt more revisions; if a revision is unacceptable, it is easy enough to recover the original. Second, the ease of producing an elegant final version with the printer has increased student interest in sharing writing with other students and with parents, and has led also to the publication of a monthly collection of stories and essays (which a student panel selects from those submitted). In addition, because it is so easy to get access to each other's work, Bob has decided to give credit this semester to a student who revises *any* story written by any student in the class. Students can submit the original and a new revision, which may differ considerably from the original author's own revision!

But the particular advantage for Bob is the improvement the computer has made in his system for assessing students' progress. Since work-in-progress is printed out frequently from the student's diskette, Bob has a file of first drafts and revisions, complete with dates, for each student. Before a meeting with a parent or a conference with a student, he can review the student's file, noting the number and kinds of revisions the student makes now, as compared with six months ago.

Now, students' work is always available; first drafts don't get eaten by the cat (or if they do, a fresh copy can be printed out easily); students themselves can see the progress from their early to their later work. As the system became established, some students realized that a deliberately horrible first draft gave them a chance to earn revision points easily. Bob soon dropped the "automatic 50 percent credit" rule; using evidence from the students' own files of writing samples, he raised his own and their expectations for acceptable work.

In Bob Bender's classroom, as a result of using the computer, the assessment criteria are firmly linked to the curriculum itself—perhaps an ideal example of a computer assessment tool that is meeting the specific needs of a particular group of students and is flexible enough to change when those needs change.

Spelling: Meeting Individual Needs

The second example of teacher-made computer assessment systems involves a spelling program in a fourth-grade classroom. Sally Taylor has always tried to individualize spelling in her classroom. Last year she had five different spelling groups and found it very difficult to keep up with providing different lists, different activities, and different tests for each group each week. Even within those five groups, the capacity of children for learning words was not the same. Since her goal was not just to get through a list a week, but to help children learn words that would be useful to them, she wanted to individualize even more. Ideally, she would have liked each child to have his or her own words, including some specified by the teacher, some chosen by the child, and some pulled from the child's writing. These words could be kept, learned, reviewed, used, and gradually developed into a personal dictionary of known words. Sally had tried pieces of this system with some children, but found it difficult to monitor the system and assess progress; she was unable to use it with the entire class.

While watching another teacher use a commercial computer spelling program with her class, Sally became interested in how her ideas might be realized by implementing them on the computer. With the help of a high school student intern with programming experience, she developed a program that allows each child to enter an individual list of spelling words each week (she chooses the words for some children; some choose their own; for some she chooses part of their lists). It then allows them to practice their words using four different kinds of exercises: scrambled words, words with letters missing, sentences (which the children themselves have to enter), and associations. (These are "mini-definitions," which the children also have to provide. An association for "fiery" might be "hot" or "burn"—whatever makes sense to a particular child.) When the child feels he or she is ready, but no later than Thursday, the child can ask the program for a test using the sentences or associations, or can have another child be the tester, using a computer printout of the week's words. The child can take the test as many times as needed. On Friday, the words that have been learned, with their associations, are added to a list that is the child's personal dictionary (also maintained on the computer) and any words not learned are saved for the following week's list. As an ongoing review, every few weeks, children are paired for a test of previously learned words, chosen randomly from each child's dictionary. Children also use their dictionaries to check words they need to use as they do their own writing.

Sally is very pleased with this system. It allows her to provide appropriate work for Cerise, who needs to learn words in patterns, Ben, who works on only two words each week, and Toby, who can spell anything and uses her dictionary as a way of collecting interesting words. Sally has a complete record of each child's work to date and can individualize as much as she or the children want without spending unrealistic amounts of time on the process.

UNDERSTANDING CHILDREN'S THINKING

One of the criticisms of many testing procedures is that they give us no information about the process that a child has used in coming to choose a particular answer. If the primary purpose of an assessment procedure is to provide information about the child's understanding, then how the child arrived at an answer is just as important as which answer was chosen. If a child chooses an incorrect answer, we want to know why. Does the child have a perfectly good reason for thinking that an answer other than the one the test-maker had in mind is correct? Is she guessing? Could he read the question? Did she misunderstand the directions? Is the lack of understanding partial or total? Does the incorrect answer reflect a profound lack of knowledge, a careless error, surprisingly sophisticated thinking, or a lapse of memory? The answers to these questions are considerably more important in the process of assessment and prescription than the simple tally of right and wrong answers.

Error Analysis

Error analysis is one process that teachers have long used to uncover children's misunderstandings in arithmetic. If it is not possible to work individually with every child and ask them to explain how they arrived at a particular answer, the patterns of errors that appear on their written work may provide clues to their thinking. Common errors in particular processes become familiar to teachers and can, therefore, be spotted quickly. For instance, look at the following two examples of a child's work and see if you can figure out what consistent, but incorrect, process each uses.

$$
\begin{array}{llll}
\text{a)} & 65 & 72 & 50 \\
& -43 & -48 & -25 \\
\hline
& 22 & 36 & 35 \\
\end{array}
$$

b)   48   135   172
   +34   +258   +245
    82   483   417

The work of the first child is fairly easy to decipher. This child subtracts the smaller digit from the larger digit in each column, no matter which is in the top number. Of course, although discovering this pattern gives a clue as to what the child is doing, there is much more for the teacher to discover about why the child is making this error. The second child's work is a bit more difficult to understand. Perhaps the error in the second problem is due to a momentary lapse of attention. On the other hand, perhaps there really is a consistent error here: the child may always place the "carried" number over the column furthest to the left. Additional problems or a discussion with the child would help clarify the initial impression.

Cataloguing common children's errors has been done by many educators, and a computer program has been written by John Seeley Brown which scans a set of children's problems, matches the incorrect problems in the set against a bank of information about incorrect processes, and produces a listing of the error patterns that would best account for the incorrect problem. However, the program requires a powerful computer system and, even then, it is not always successful in identifying the incorrect process a child may use. A computer program that analyzes errors in this way may give important initial information about a child's incorrect rules for solving a particular type of problem. However, it cannot describe the underlying misunderstanding that is at the root of the consistent error.

Observing Children Solving Problems

In the case of error patterns in computation, the teacher can often clear up what is going on by having the child give a verbal explanation or a demonstration with concrete materials of the process used. However, some children are unwilling or unable to explain their approach, particularly when they know it is yielding wrong answers. Many of us have had the experience of sitting in a lecture that we cannot quite follow; when the lecturer asks for questions, we are full of confusion but are unable to formulate a question that will help us understand. Similarly, a child who is confused and cannot explain why may shrug and fall back on the perennial, "I don't get it"; a child who is fearful of failing may lapse into silence; a child who is non-vocal, bilingual, or simply uncomfortable with oral expression may be unable to find the words to communicate what he or she is thinking; a child with impaired hand use may be unable to show what is intended by using objects or gesture; a child who does not hear normally may not be able to understand a question or provide an explanation.

Assessment that tells us most about the child's understanding often relies neither on conversation nor on the final product the child produces, but on observation of the process the child uses. Creating problem situations in which the process of the child's thinking becomes transparent through his or her actions thus becomes one of the major tasks in the assessment process, if the results are to be used in understanding the individual child's needs and in planning an appropriate curriculum.

Like many other educational activities, some of the computer activities described in Chapters 3 and 4 of this book lend themselves as much to assessment as to use for learning. For instance, watching a child's strategy in the game *Darts* (described in Chapter 4) could provide a great deal of information about the child's understanding of fractions.

Assessment of reasoning and problem-solving skills is particularly difficult and time-consuming, requiring careful observation of a large amount of complex information rather than scoring answers that are either correct or incorrect. Some teachers already consistently use observation of their students as a means of gaining information, not about the final outcome of their work, but about the difficulties and needs they have while they are engaged in a particular task. However, the cost in terms of time spent in such activity is often exorbitant.

The computer is exceptionally well suited to the creation of problem situations that are inaccessible in the real world, either in general or for a particular child, and that allow convenient observation of the process the child uses to solve problems in that context. Here are two examples of children working in such a situation.

Consider a group of children playing *Hunt the Wumpus* (see Publications in the Resources Section), a popular computer simulation in which the players are in a maze of interconnected rooms. Players are given clues about where the Wumpus is hiding as well as about other nearby dangers, such as pits and bats. A skillful player can gradually develop a picture of the maze (which changes for each game) and use the clues to locate and subdue the Wumpus before getting eaten or meeting some other disaster. One of many possible objectives for this game from a teacher's point of view might be to assess students' ability to record and organize information, an essential problem-solving skill.

Two fifth graders, Susie and Greg, approach the game quite differently, demonstrating both their strengths and difficulties. After a few moves, Susie begins drawing a map of the rooms, noting information given in the clues, and carefully moving only into safe rooms. She wins the first game, largely because of her care in arranging important information; however, she has difficulty during the second game because she continues to use previously recorded information from the first game instead of constructing a new model from new information. Greg does not record any information during the first game and is quickly vanquished by the Wumpus; during the second game he still does no writing but begins to rehearse what he knows orally as he works, "Let's see, rooms 3 and 11 are safe and 5 has bats. . . . Ok, 3, 11, and 6 are safe, 5 has bats, the Wumpus is near 6." He doesn't remember which room is connected to which others, and he sometimes forgets a piece or two of information, but his lists very often work well. Occasionally, he takes a calculated risk, moving into a room he knows nothing about; sometimes this is disasterous; sometimes, he gains a new piece of information.

Both of these children have strengths to build on and the need to develop or expand their repertoire of approaches further. The observant teacher can gain a great deal of useful information by watching these children at work, not only about the methods they use to keep track of information, but also about their styles of solving problems. Together with other information about these children (e.g., Greg's resistance to writing down anything in other contexts and Susie's rigid, and often incorrect, application of rules in solving mathematics problems), observations about their behavior in this context suggest curriculum strategies, including use of *Wumpus* as a learning tool.

At the Massachusetts Institute of Technology, Dr. Sylvia Weir's group is using the Logo language to investigate children's understanding of space and number. Here again, the computer is seen as a tool that makes the process of children's thinking more available to the adult observer as children solve problems involving such skills as estimation of distance and angle, counting, matching, determining direction, and mental rotation.

Ned, a non-vocal, 6-year-old, with cerebral palsy, began to work with a robot turtle that he controlled by using a buttonbox (six large buttons in a plastic box). His use of the buttons to direct the turtle to move forward or back, to turn right or left seemed random during the first session. He had a great deal of difficulty moving his hand onto a button, pushing the button, and then lifting his hand off the button to terminate an action. Observers could not tell if the random nature of his actions had to do with his motor difficulty, his desire to explore the possibilities, or lack of understanding about the nature of the activity. However, by the third session he had figured out how to manage the buttons well enough to engage in a game of sending the turtle into "caves" (under chairs), using right and left turns accurately no matter what the orientation of the turtle in relation to himself—a feat for any child of this age. He showed his engagement in the play in many ways, including his use of the turtle's "eyes" (small light bulbs), turning them on just before entering a cave and off upon leaving. Teachers were able to learn a great deal about Ned by watching him work with the turtle, such as his appropriate choices of right and left, his ability to engage in creative play, and his long attention span during this activity (despite the hard work of motor control in which he was continuously engaged).

Logo can provide an environment in which children devise and solve their own problems. Teachers and others can then use it to look at the approaches and techniques, both successful and unsuccessful, that students use in the problem-solving process. Educators need to be alert for other similar tools and software that can be used for both assessment and learning, even if they are not specifically designed for assessment.

The least interesting computer assessment may be precisely those pieces of software that are designed specifically for that purpose. For instance, many drill and practice programs include tests or, at least, criteria for advancing from one level to the next. While these are certainly not useless, and may be particularly helpful for some children who are frightened by the usual pencil-and-paper test formats, they do not offer the educator much more than the pre- and post-test provided with most textbooks. They provide information on limited-goal objectives, not on more complex cognitive skills embedded in a situation that holds the attention of the child. In contrast, uses of the computer that open windows into children's thinking hold great promise for assisting in developing a complete and sufficiently complex description of the child's understanding.

WHY DO I HAVE TO TAKE THIS STUPID TEST?: ASSESSMENT AS COMMUNICATION

The previous sections have dealt with two of the six aspects of assessment: the relationship of the assessment tool to the curriculum and the amount of probing into the process of children's thinking which occurs during assessment. Three of the other aspects have to do with communication and attitude—how the child interacts with the assessment situation. They are: 1) the way in which directions, questions, or problems are communicated to the learner; 2) the method of student response; and 3) the attitude of the learner towards the test. The results of an assessment can be partially or completely invalidated because of limitations created by any of these factors, whether the tester is aware of them or not.

Elena is a 15-year-old bilingual student who recently moved to the United States from Puerto Rico. She is taking the vocabulary portion of an achievement test. The test is in English, her second language. Her teacher notices that after the first three items, she is moving her pencil rapidly down the answer sheet, apparently selecting answers at random without even looking at the questions.

Jeff is twelve and is repeating fifth grade. Observed doing a page of mathematics problems, he is chewing his pencil, twisting in his seat, and has already made two holes in his paper by his tense erasing.

Ned is the 6-year-old child with cerebral palsy mentioned in the previous section. His teacher is trying to determine how much he knows about numbers—how high he can count, which numerals he recognizes, whether he can match numerals to quantity, whether he understands the concepts of more and less. Ned cannot articulate words or manipulate objects, so his teacher is asking him questions about groups of objects and numeral cards which she manipulates; he points to "yes" or "no" on a large board in front of him. While she suspects he understands much of what she is asking, he points erratically and ambiguously.

None of these children may be having difficulty with the intellectual content of the test, although their performances seem to indicate lack of knowledge. In fact, Elena does not understand the English directions, although she would do well on a similar test in Spanish. Jeff has much less trouble with the actual mathematical content of his worksheet than with isolating one problem at a time on the page and writing the numerals he means to write down in the correct order. Ned's teacher thinks he is refusing to play her game. Perhaps he is uninterested, or tired of being probed; perhaps he is frustrated by the disproportionate effort required to show a very little of what he understands.

All of these children feel that the testing situation is out of their control. They are scared, confused, or feel that the test is unfair. If you are not going to have the opportunity to show what you know, if you are sure that the outcome is going to make you look stupid, why try? "Just do the best you can" is advice that makes little sense to many children—and adults—who find themselves being assessed. The lack of feedback in most testing situations, inclusion of items designed to be too difficult, lack of time, confusing format or directions, the absence of flexibility in modes of response, and previous experiences of failure have often convinced those being tested of their inadequacies before they even start.

Computers and Testing

While the computer is hardly a panacea for these problems, it can assist some children in establishing control over the testing situation and in overcoming communication barriers. Here are some of the ways the computer can be valuable:

- Some children seem to find the computer less judgmental than human testers. Children who are oversensitive to adult cues may find evidence of disapproval even where it does not exist. Some of these same children are more comfortable interacting with a computer; it feels more private and more within their control.
- The format of tests causes many children to have difficulty: how questions are spaced, the form in which the answer must be recorded, the amount of reading, the nature of the directions may all obscure the intent of the test. While educators can devise their own formats

and modify assessment tools to meet individual needs in local situations, they cannot always control these factors for all the tests with which a student may have to cope during his or her education. What they *can* do is to provide the student with opportunities for practice with various formats and directions. The computer is an ideal tool for giving students such practice, at their own pace, with immediately available feedback, and no immediate consequences for failure. For example, a program which produces problems in the inequality format used on the Stanford Achievement mathematical computation subtest (6 + 4 ___ 2 × 6: choose <, >, or = to go in the blank) gives a child useful experience with problems in this format, one at a time, accompanied by feedback on each problem and perhaps even by a pictorial explanation of how to solve the incorrectly answered problem. As discussed in previous chapters, the computer can be a prosthetic that allows clearer and easier communication by the student. This has particular significance in the testing situation, not only for children who lack speech or motor control, but also for children who have learning problems which impair their ability to deal with written symbols. Anxiety about test-taking is often linked to inability to communicate. Being unable to tell, write, or show what one knows is, at best, frustrating; at worst, it may convince not only the teacher but also the student that she or he must be inadequate.

How might use of the computer help Elena, Jeff, and Ned to overcome the problems they have in testing situations? Of the three, Elena probably has least need for a computer; rather, she needs a more appropriate testing instrument that does not simply remind her how little English she can read. Indeed, not all problems are solved by computer application. However, both Jeff and Ned can be helped through their difficulties by using a computer.

Jeff has a serious learning disability that manifests itself in the difficulty he has recognizing and reproducing written symbols. Although he reads, with some struggle, his writing is tortuous. Jeff began to use the computer to learn Logo, typing in commands to create elaborate pictures of his own devising. Jeff has enormous trouble finding the keys on the keyboard and often mistypes even familiar commands, but the ease with which mistakes can be corrected, by using the rubout key, is invaluable to him. He is willing to grapple with what seems to him to be the disorder of the keyboard because correction is so easy. He is now willing to attempt some longer creative writing and takes all tests on the computer. Because some of the communication difficulty has been removed, or seems, at least, more under his control, his attitude towards testing has improved. The results on tests he takes now give a more accurate reflection of his abilities, rather than simply showing that he hates tests.

For Ned and similar physically handicapped children, transferring assessment tasks directly to the computer may be an important step. It is clear from his work with the turtle that he can learn to control an input device such as a buttonbox. For other children, joysticks, light pens, or voice input might be appropriate. Tasks such as counting, matching numerals to quantity or quantity to numerals, seriation, and classification can be implemented on the computer for both assessment and instruction. With the barrier of inadequate motor and speech control partially removed, Ned is more willing to put effort into showing his teachers what he knows. Gaining communication control is tightly connected for Ned with motivation, attention, and interest.

INFORMATION OVERLOAD: COMPUTERS AS MANAGERS

Use of the best possible assessment tools is not enough. In order for the information gained through assessment to be valuable, it must be recorded in a form that is useful to teachers, parents, and students; it must be accessible when needed; it must be updated and passed on to those who will be responsible for the student's education in the future.

While the intention of the Education of All Handicapped Children Act, P.L. 94-142, is to provide individually appropriate education for each child, the paperwork that it requires consumes many hours of teacher time. Individualized Educational Plans (IEPs) allow educators and parents to agree on appropriate, clear objectives for each child and to make sure that no one falls between the cracks. However, keeping records that reflect the complex totality of a child's progress through school is a task that has caused many teachers anguish. While most people would agree that a report card listing letter grades hardly reflects the characteristics of a student's learning experience, the thorough record-keeping of a student's cognitive, social, and emotional growth is a full-time job in itself.

Computers are good at keeping track of large amounts of information, allowing it to be changed, deleted, or updated easily, and keeping it accessible at all times. Computers can be used simply to keep track of objective information (addresses, phone numbers, hours/week of special services, etc.), to record and update IEPs, or to keep more anecdotal records. Such systems range from those that are simple, home-made, and relatively cost-free to complex, commercially available systems.

Keeping Track of Observed Behaviors

Marcia Hutchens reports a small computer checklist that helps junior high school teachers keep in touch with each other about their students (see Publications in the Resources Section). Teachers wanted to keep track of what worked with difficult students from class to class and from year to year. A list of "what works" was compiled from teacher suggestions: e.g., "student enjoys being chosen to teach other students in the class," "student requires a very specific topic to write on," or "student responds to frequent communication with his parents." Teachers can now use the computer to select from a checklist of about twenty such items those that apply to a particular student (teachers can also add their own); other teachers of the same student can then refer easily to this list and add comments of their own. Because the information is all positive, Hutchens notes, the system helps develop a supportive attitude toward difficult students. Teachers also think the system is "fun" and approachable, so are likely to use it and thus share important information.

This system could easily be expanded to include longer anecdotal records. Teachers who have tried fairly elaborate systems of record keeping using notebooks, note cards, and daily or weekly recording have often found their own systems unwieldy. Any computer based text-editing system could be used to keep a file for each child, perhaps with appropriate categories—mathematics, language, peer interaction, art, and so forth, depending on need. Anytime during the day, at the end of the day, or once a week, the teacher could call up a child's file and type in quick notes in the appropriate category, e.g.:

Math 11/14 While picking up blocks today, counted to 18 without skipping numbers.

Conflicts 11/15 Kim and Alicia once again in fight on playground. Kim continued to try to keep the fight going when Alicia ready to accept intervention.

Language 11/16 Perry said clearly "more juice" at lunch.

These notes are reminders of what has happened over several weeks or months and can provide concrete examples of learning and change for use in conferences with parents and students. Others who interact with the child can also be encouraged to use the system, including student teachers and specialists.

IEPs

Many computerized systems have been developed, independently by schools and commercially, to manage the IEPs of students with special needs. Some of these programs do not deal with individual educational goals, but do keep track of name, address, phone, dates of evaluations, scores, services received, and other objective information. These programs can find and print information for a particular pupil, give lists of pupils in particular categories (e.g., all pupils receiving speech therapy), print mailing labels, and so on. Other programs include a catalog of objectives from which IEPs can be developed. Some schools develop their own programs which can run on a single microcomputer. Others purchase commercial systems or contract with companies which receive IEP information sheets from the schools and process them on their own larger computers.

While the management possibilities are exciting and are a sensible way to use the computer's capacity to store data efficiently, educators will need to be wary of systems that are not appropriate for their school's needs and objectives. We do not want to be subtly influenced by a commercial system's list of objectives; rather we want a system we use or develop to be flexible and responsive to our curriculum and our students.

Kirk Wilson's Teacher Planning System (see Software in the Resources Section), for example, is an interactive system that is capable of storing a vast amount of information for each student. Yet, the categories of information, format of information, what kinds of reports can be printed out, and who has access to which parts of a student's record are all determined by the users of the system in the local situation. Goals for an individual student can be selected from a library of goals by teachers to match particular needs. The system is easy to use without knowing anything about programming or even about computers. The best possible outcome of use of such systems will be to maximize efficiency, accuracy, and usefulness of necessary record-keeping, while freeing teacher, counselor, and administrator time to deal with substantive educational issues.

CONCLUSIONS

The examples in this chapter illustrate uses of computers for assessment and management designed by educators to meet the needs in their own situations. Computers can support flexible assessment that is sensitive to the progress and abilities of individual children. They can help the overburdened educator or administrator keep track of large amounts of information and access any piece of it easily. They can allow children to have some control over their own testing. They can act as a communication prosthetic in assessment situations. They can provide simulated problem situations in which teachers can observe the process of children's thinking. They can allow teachers to design assessment/curriculum tools that are tailored to fit their own classroom situation.

But computers cannot make decisions about the quality of an assessment tool nor can they interpret the results of testing. Moreover, if used inappropriately, computer-based testing and assessment can result in *less* attention to individual learning and communication needs. Nevertheless, if educators, parents, and students are all involved in creating, monitoring, and modifying assessment tools, the power for developing new and exciting models for assessment is in our hands.

Getting Started 7

You do not have to know what bit or byte or RAM or ROM mean to work successfully and happily with a computer. You do not have to know binary. You do not have to excel in mathematics or science. You do not have to be mechanically inclined. Like the family car, a computer requires some skills to operate, but also like your family car, those skills are acquirable without knowing or caring much about the terminology, construction, or repair of internal combustion engines, brakes, or transmissions. Still, a good car owner's manual tells you enough about the technology so that you can be at home with it, care for it, and use it safely. You should have that kind of information about your computer equipment. The fastest way to gain the absolutely necessary knowledge is by asking someone to show you. If you prefer to read about it, there are several good sources to which you can turn. The first chapter or so of the book that accompanies your computer will provide you with the specifics of the care and maintenance of that particular machine. Chapter 3 in the *Practical Guide to Computers in Education* (see Publications in the Resources Section) presents a thorough and readable introduction to some more general concepts and terminology associated with computers in classrooms. For other introductory books on the subject, see the Resources Section of this book and of the *Practical Guide*.

BECOMING THE SCHOOL COMPUTER EXPERT

If it is your personal interest to become your school's expert on computers, feel encouraged. Most teenagers who become computer wizards do not initially bring any background in mathematics, science, or technology beyond your own. It is interest, courage, opportunity to learn, and plenty of hands-on work/play that let them become proficient. There is no standard order in which to learn about computers. In particular, you do not have to learn BASIC before you do anything else with your computer.

If it is *not* your personal interest to become a computer expert, then don't. On a day-to-day basis, you won't need that knowledge. When you do need special expertise for designing or programming something completely new, there are several sources of potential help available to you.

What do you do and how do you do it? To whom can you turn for guidance and support? Can you implement directly the work done by other school systems? The school systems that pioneered the use of computers in education are important to learn from, but are rarely easy models to follow. A pioneer is one who got there early and blazed the trail. But computer technology is changing with amazing rapidity. If that trial was blazed on foot, and you need a moving van, the trail itself is of little use. You may benefit from what others have learned about the lay of the land, but you will have to build your own road. For example, if a particular computer trail was blazed before 1976, it did not use microcomputers; for all practical purposes, there weren't any. And economic realities are also changing rapidly: you can buy that moving van today for the money that would have bought only a bicycle eight years ago. For over twenty years, computing has reduced in cost (or, equivalently, increased in power) by a factor of two every two years. That rate is predicted to continue or even speed up. School systems with already established programs began their work with poorer technology than is available today; that technology continues to influence the way their programs work today, determines what kinds of successes they have had in the past, and it certainly colors the way they think about computer use.

Old Thinking about Using Computers

"Old thinking" has already limited schools' visions of what they can do. As recently as 1983, a national survey reported by Marc Tucker determined that of the many ways in which computers may be used in schools, rarely does one see examples of any but the two most conventional uses: 1) computer-aided instruction (CAI) which meant, for the most part, drill and practice, and 2) "computer literacy" which, in most places, consisted of the teaching of BASIC. For the first, the rationale was cost efficiency; for the second, a combination of goals, foremost among which were general preparation for life in a technological world and specific preparation for the job market. The cost efficiency of CAI is, at best, debatable. There are studies that show it to be effective, but peer teaching is also shown to be effective, and it is cheaper. This is not an argument against CAI; it is merely to suggest that cutting instructional costs is not by itself a good rationale for installing computers.

The job preparation argument is even more spurious. There are roughly 40 million students in our schools, and about 20 thousand computer programming jobs—two thousand students for every programming job! Even if the number of such jobs were to increase tenfold—and the best current wisdom predicts exactly the opposite, a dramatic decrease in the number of programmers in the future—it is clear that preparation for the job market is not an adequate justification for a universally taught course. Nor is it necessary to learn any programming language as a general preparation for life. Predictions are that by 1990 half of the workers in the United States wil be using computers, but most of the uses will not require any more technical knowledge than is now required for the driving of a car. Even now, there is plenty you can do with a computer—for example, word processing—without knowing any computer languages at all.

Are we arguing against the use of computers in education? Certainly not! Merely that there is no *universal* need, no single body of knowledge or skills that all students must have for survival, psychological or economic. For some students, computers represent, in themselves, a fascinating subject matter. For others, they represent access to interesting activities. For still others, they may simply be a change from the routine. For the teacher planning on using computers, this is a liberating reality. One school painstakingly duplicated their favorite workbook lessons on computers, right down to the graphical detail of representing on the screen the spiral binding those workbooks had. Some students preferred the pizzazz of the machines, and others preferred the paper workbooks. For that school, offering their students that choice, alone, made the computer worthwhile.

The point is that you and your students remain the final arbiters of what is useful to you. There is no threat "out there," no technological authority to whom you must defer in planning the activities of your classroom. There is always new knowledge to have, and computers offer new ways of knowing, but you must still decide yourself how best to help your students and you can use the same wisdom you used yesterday. The difference is that today you have some more tools with which to provide that help; more ways to apply that wisdom.

Another rationale given for the teaching of programming is that it directly enhances education—not as a preparation for some future job, but as a tool that sharpens thinking skills in the present. While it is still too early for research to have unambiguously verified this claim, there is considerable face validity and a plethora of testimonials to support it. For this purpose—teaching students to program as a technique to help them organize their thinking—the choice of language and activity becomes significant. When BASIC was first designed, it opened up computer access to a wider audience because it was cheap to run and gave students fast feedback on their programming errors. Today, there are other languages designed for students. Logo, for one, is not only easier to learn, but also more powerful and far better for encouraging good thinking habits. Thus, if you want to teach your students to program, whether as young beginners or as advanced high school students, Logo is likely to be a better choice than BASIC. Clearly, if your reason for teaching programming is essentially as vocational training, then the language chosen should reflect specific jobs that are expected to be available to your student.

In this chapter, we will suggest some techniques and point you to other resources to get you started and help you find the further support that you will need. Yes, you *will* most likely need further support. Fortunately, it can be found.

Getting Help

First, we suggest that you make a list. Use this book as a resource to write up the kinds of computer activities that you might immediately want in your educational environment, and the kinds of software and devices that these activities seem to require. Whether that activity is described in the book or is your own idea inspired by something you have read, keep track of the page on which the source can be found. It can be useful for helping you flesh out the idea later in the planning stage.

Then, make a second list of what you envision that you might be using by the end of the year. Keep in mind your students' current interests, but allow yourself to dream a little about their potentials. The problem at this phase of setting up a program is often one of underestimation. Pick at least some activities that are either at or just above what you think are the students' upper fringes of capability, as well as some activities that stretch beyond what now seems reasonably possible. When a student gains power in an area of interest, the growth can be staggering.

If you do not yet have computers and must first decide which machines to buy, you will have very different questions than if you already have computers and are trying to augment their use in your classroom. In either case, decisions should be based both on familiarity with the available technology and on knowledge of and sensitivity to educational issues. It is not necessary that the same person have expertise in both areas, but it is very necessary that the two experts communicate well. There are lots of books that claim they will make you enough of an expert to buy and understand your computer system. If you want a fairly conventional system, capable of running commercially available games, educational programs, and other packages, this may be true. But the key to success in the special education classroom is flexibility—non-standard applications. If you want to design a rich, flexible, and powerful computer environment like some that are described in this book, one or two introductory how-to manuals will not suffice. Find someone you can work with who knows the technical details. There are many sources to tap, and you may find that the combined help of more than one person is required. Computer hobbyists, old or young, generally have a great wealth of knowledge and experience, and tend to be very willing to share not only their knowledge, but their time as well.

Often, local high school students are good sources of programming support. Sometimes, they may volunteer their help, and sometimes it may be more appropriate to arrange for them to do work in return for credit or payment. Some communities have successfully put out programming contracts for high school students to bid on. A detailed description of the project is posted, and students, in return, write proposals for the implementation of the work including estimates of the length of time it will take. Past performance (both reliability of the estimate and quality of the finished work) as well as the current bid are taken into account when selecting the student to hire. High school students may need direction and supervision but can often supply considerable technological expertise that you don't have time to develop. Students who own their own computers may even be relatively reliable sources of the latest information on what hardware devices are available.

University students and professionals are also potential help. Local universities may be very excited about having a real-world, practical, and innovative project for their students to work on, either as a specific complement to their classwork (e.g., a term project) or as related, paid outside employment. Even where distance precludes any regular involvement, sometimes the opportunity to help set up an interesting demonstration project is attractive enough to warrant the initial contacts. You get some support, and the university acquires an interesting educational experiment to watch. It is also worth knowing that software developed under public support (for example, in a university project supported by a grant from the government), is generally available free or for nominal cost to cover postage and media.

Local industries with computer-knowledgeable personnel may also provide help, if asked. Public relations, tax status, and employee goodwill all stand to benefit when industry donates the time of interested employees to worthy community causes. Even the employees of a local computer store may be willing to give (or sell) help outside of their normal hours.

In some communities, the parents of your students may be a very fertile source of contacts. In others, you will be more dependent on sources to which you can write.

Find people who can be of specific help in what you are doing and form a development team. If you are selecting computers, then at least one member of the team should be familiar enough with the details of different machines to help you make that choice. If you need some special device, you want people who, at least collectively, are familiar with a wide range of products and can help you decide whether you are better off buying it or having it made. If you are designing software, someone in the group should be able to program well in a language suitable for that purpose. If you are teaching programming to your students, still other skills may be needed.

If you are not a technical expert, and if your technical colleagues are not experts in education, invest some time in sharing ideas and expertise. Neither of you is going to do the other's job, but this sharing helps communication and helps to establish the working relationship you will need. An efficient way to begin is to have your technical experts read parts of this book—at least Chapter 1 and any other sections that are relevant to your list. Other useful information for them includes a list of the computer equipment and software that you currently own, and the resources section of this book.

Be sure to clarify the task. You are not asking your technical experts to design educational programs in a vacuum; they are not the educators. You and the computer expert need to work together as a team. You are an expert in your field and, now that you've read this book, you know the kinds of things that are possible to do with computers. You have already made some selections among possible alternatives, and have begun to plan an educational environment. With technical support, you and others will implement that plan. You are asking for help in locating the best currently available hardware and software that meet the specifications on your list, planning any special adaptations that will be needed, and setting up the system when the components have been bought, programmed, or built.

Technical colleagues are worth cultivating even after initial tasks are finished. While they provide service to you, they are also being educated in some of the areas of your specialization. As a result, they become even more valuable resources in the future, both to you and to other teachers. The work that you do is potentially of great value to other teachers, schools, or students. Consider joint publications of articles describing that work, or even commercial publication of the products themselves.

The sources of these technical contacts may supply additional valuable help. Once your system is set up and running, you are still likely to need technical support to provide new activities and revamp old ones. Local university students might be willing to provide help in teaching programming or other technical skills to those in your classroom who are interested, if you do not have the expertise or time to do that task yourself.

You, meanwhile, might benefit by reading Chapters 4 and 5 of the *Practical Guide to Computers in Education*. These chapters raise some of the issues related to hardware and software selection, and may help you translate your plans and educational concerns into technical requirements.

We will not repeat here all of the information that you can gain from other sources, but must emphasize a general caveat: when buying computer hardware or software think in terms of the capabilities you want, not in terms of the standard terminology used to compare machines. Bigger memory size for example, often does matter, but some machines provide capabilities in a compact way that on another machine would demand more memory. It is the capability, not the size, that is critical. If bigger is not always better, it is also important to know that "bigger" is not always even bigger! Not all of the memory that a computer has is available to the user. Again, it is not the number that counts, but what the machine is capable of doing for you. Use your own expertise efficiently. Design the *educational* environment that you want, listing its tasks and capabilities specifically—no one else can do that. Then work closely with someone you trust who can help you translate those desires into technical requirements.

CLASSROOM ISSUES

Most of the answers about safety and appropriateness of computer technology with children are the same whether the child is classified as "special" or not. However, the questions often are not. Will a child who drools be safe with the computer? Will a child whose movements are uncontrolled and erratic be safe? What about a child prone to seizures? Are there special modifications that can be made for a child who cannot type? Are there any modifications that need to be made for a child who cannot hear?

Preserving the Equipment

Like any other piece of equipment, computers can be mishandled, but they are less sensitive than many other common classroom technologies. Short of spilling Coke in them or wedging pencils in the keys, the **keyboard** and **processor** parts are relatively invulnerable. Misunderstanding a program and typing the wrong keys, even tripping over the power cord and accidentally pulling it out, cannot damage the computer. The worst effect of such a mishap is the loss of some information, and in the most likely case, even that will not be a large or irrecoverable loss.

The **disk drive** and **disks** are a bit more touchy. The information on disks is stored as magnetic marks, somewhat the way audio or video tape is recorded. As a result, lying too close to other magnetic equipment (any telephone, most screwdrivers) can interfere with the contents of the disk, erasing or garbling the information. Likewise, dirt on the disk's surface—or even the oils from one's fingers—can shorten the disk's life span. Crumpling the disk does it in very quickly. Even so, disks are fairly robust. If a telephone rests on top of the disk, and rings, the magnet that rings the phone bell may cause some problems, but there is no danger to a box of disks sitting even an inch away from the phone.

The disk drive should have nothing but disks inserted in it. The disks, both for their sake and for that of the drive, should be put in considerately. Some drives are (mildly) sensitive to being operated while empty. But, again, recall that the manufacturers of these products knew full well that they were building equipment that would be handled by people, some of them young and exuberant, who knew little or nothing about computers.

Computers are relatively hardy, but to know what they can and cannot endure, get to know your own equipment well and ask others who have used similar equipment.

Preserving the People

Safety for the person using the computer is also well assured. No electrical equipment should be treated carelessly, but computers are among the safest electrical devices for a number of reasons. In general, computers use very low voltages. The wires that send signals from the computer to the TV or printer carry less voltage than is used in a transistor radio. Small computers generate far less heat than a TV does. Even dropping the computer is less likely to cause harm either to the equipment or to people than dropping a TV. By far, the TV is the more hazardous piece of equipment. Of course, the various wires that run in and out of computers should be arranged in such a fashion that people are not likely to become tangled in them! Dangling wires can cause accidents even when no electricity is involved.

Eyestrain may be a consideration for students who spend large amounts of time with the computer. However, by adjusting the contrast on the screen and the lighting in the surrounding work area, eye strain may be kept to a minumum.

Planning the Room

In all settings, a well-planned physical layout of the equipment improves the work environment not only for safety to people and equipment, but for productivity and appearance, as well. A location should be selected that minimizes the need for and contact with trailing wires. Even in the set up of the equipment itself, criss-crossing tangles of wires not only make it harder to track problems with the equipment when they arise, but also tend to be distracting and unattractive when they are visible. As the technology changes, these considerations will change as well. There are now several truly portable computers. They operate off batteries and contain their own tiny displays, therefore eliminating the need for dangling wires. These tiny computers also make it possible to develop wheelchair-mounted aids that move with their user.

Traffic patterns in the room are important, too, but the choices are less simple. A location near a door which has constant traffic may be distracting and hazardous. Also, machines packed too closely together pose not only a problem of distraction, but may also make it difficult for students to spread out papers that they may need or want to work with. On the other hand, a location that is too private may have social disadvantages, depending on the nature of the classroom and the purposes for which the computer will be used. The individual relationship between child and machine may be extraordinarily liberating and healthful for one child, attractive but unhealthful for another, and just plain lonely for a third. Some computer-supported activities benefit greatly from social involvement, the chance to show a friend or work together. Others are valuable because they allow the child a chance to work with privacy and anonymity. If it is feasible to create some variety in computer environments, both you and your students will have more opportunity to choose the atmosphere that best suits the moment.

Mounting or Housing the Computer

Most often it is adequate just to set the equipment out on a table, but sometimes circumstances make it worth the trouble to find a way of mounting and housing it. Mounting can help to prevent unwanted moving of the equipment. Sometimes, mounting is desirable for precisely the opposite reason, to increase portability by binding together several pieces of equipment that are most easily moved as a unit. Proper mounting can also provide a protective shell for the equipment, reducing the wear and damage to it that results from normal buffeting.

Protective mountings, if any, should consider the possible hazards to children who are likely to make uncontrolled movements, whether they result from behavioral or motoric problems. Sometimes the problems are difficult to foresee. For example, a special board had been built to house some computer equipment that served Jan as a communication and educational aid. The board was designed to fit directly over Jan's wheelchair as a lap tray. A plastic overlay protected it from accidental spills and occasional drooling. Jan communicated by touching various locations on the board with his fingers, mostly using his right hand. Because of his severe spasticity and athetosis, however, he continually rubbed his elbow against the rear edge of the board resulting in injury to him. Eventually the board had to be redesigned. Both the shape of the top surface and the material from which it was made had to be changed.

For some special education applications, portability is essential. A student who depends on the computer for personal communication may need essentially constant access to the machine. The devices that Jan and Eric used were part of the standard equipment they needed on their wheelchair. A less handicapped student may not be able to write without the aid of the machine, but will not need it in situations which do not require writing. Some computer aids are only intended for formal educational settings and do not need to be transported at all.

Some students have no special needs concerning schedule or location. Gregory, the bright and successful dyslexic high school junior described in Chapter 5, has found that by typing his papers into the school computer's word processing program, he can edit his papers easily enough to make it practical for him to correct early drafts and perfect his work. Though he uses the computer for both English and history, and sometimes for physics, he can always do without it during class.

When designing a system, issues of portability and access may influence the decision about equipment.

Speculations

Two questions persist in coming up and for which there are not yet definitive answers.

There are continued speculations about hazards that may result from long-term exposure to radiation from video display terminals. The situation in schools differs markedly from that of business. In schools, the displays tend to be somewhat different and the exposure is considerably less. On the other hand, the persons involved are considerably younger. Is there a danger? A letter published in the *New England Journal of Medicine* cites research to support the statement that "it is generally agreed that the video display terminal is not a major source of radiation for the user." However, the authors caution that the use of color TV sets manufactured before 1970 as video screens does constitute a potentially significant radiation hazard, eight to nine times the dosage considered safe for children under 18 years of age, and suggest that newer color television receivers "decrease the possibility of excessive radiation exposure." Their computations, nevertheless, leave the question quite open. Assuming an average two hours per day viewing time at distances normal for computer use (40 cm), and the 0.5 milliroentgens per hour limit for radiation emission set by the Food and Drug Administration Bureau of Radiologic Health, a child may still exceed the National Council for Radiation Protection and Measurement's recommended maximum dosage. While there is no reason to suppose that handicapped children would be more susceptible to harm than other children in school—if, indeed, there is a risk of harm—at least some of the uses of computers that we describe may result in greater exposure than two hours per day. Is there a danger? The answer remains unclear.

Pediatric neurology departments often are asked by parents whether the blinking and flickering of computer games are likely to induce seizures in their epileptic children. At least one such department fortunately routinely answers that they do not know of any cases in which this is known to have occurred. It is known that certain frequencies and intensities of flicker are capable of inducing seizures in some people, but it seems unlikely that a stimulus like the blinking **cursor**—the spot of light that indicates where on the screen typing will next occur—would have such an effect. The blinking lights that are sometimes used in computer games for special effects, for example, to represent an explosion, or to attract the player's immediate attention are more intense, but pass quickly. Is there a danger? Again, it is not apparently attested, but the question stands. There appears to be scant literature on it.

In cases like these, incontrovertible research seems hard to come by. For years, some educators and researchers have claimed that flourescent lighting exacerbates the behavior problems of some children, and yet there remain counteropinions in the literature. The best wisdom is to observe closely and sensitively, and make a decision accordingly.

CUSTOMIZING THE COMPUTER FOR INDIVIDUAL NEEDS

One of the great virtues of the computer for special education is the extraordinary flexibility with which it can be used. The most common ways in which information is given to the computer are through the keyboard and disk drives. But there are many senses other than these conventional "eyes and ears" of the machine by which it can be instructed. Some of these devices are available as commercial products. Others are easily built from instructions to be found in the computer hobbyist magazines. It does not pay to be glib about this ease, however. By "easy," we mean it does not require hundreds of dollars of equipment and weeks of consultant time. But unless you are interested in spending time on the technology, you will want help. In each of the cases that follow, we indicate the source from which help came.

Eric

The thinking that went on in creating Eric's ironing-board-talking-device is a good illustration both of the variety of techniques and hardware available and of the criteria used for selecting among them. References to commercial sources for each of the devices mentioned below may be found under Hardware in the Resources Section of this book.

Eric needed some kind of assistance in communication and a computer-based talking and writing aid seemed ideal, but how Eric would control the computer was not clear at first. Phone calls and occasional visits with Eric's occupational therapist helped those designing his system. Eric's finger, hand, and arm control are inadequate for conventional typing, and so Eric needed some other way of conveying information to the computer. Because Eric's head control was somewhat better, it was thought that he could type using a head wand, a pointer extending from a band strapped around his head. This worked, but accuracy and speed were still a problem. A brick placed under the rear of Eric's computer gave him a better angle from which to type, but the demands were still very great. The keys were close together and demanded visual attention that Eric seemed to want to keep on the screen, the product of his typing.

Somehow the information that Eric had to choose from had to be spread out more to make it easier for him to manage. However, the more space there is between choices, the fewer choices that can be fit into an area Eric can reach. Unless the few choices that are presented lead to other choices that expand Eric's options, the aid becomes too restrictive. It was decided to use the computer to display whatever would fit within Eric's reach and then change the menu dynamically at Eric's command. The menu might first present a choice of two dozen fairly broad topics about which Eric could talk, e.g., home, school, self. Then, having selected the topic, perhaps by touching it or pointing directly to it, Eric would be presented with a new menu offering optional sentence structures, and appropriate vocabulary for that topic, always with a way of his escaping into a freer structure (e.g., the alphabet) so that he is not limited by what has already been programmed into the machine.

The question still remained: how would Eric select among these choices? Several options were considered.

Most personal computers provide sockets into which one can plug a variety of devices that resemble dials (sometimes called "**game paddles**"), manual stick-shifts (sometimes called "joysticks"), and push-buttons, all available off the shelf from the manufacturers. These are simple and inexpensive and might have been adequate. Eric could, for example, steer a little spot of light around on the screen using a joystick, and then when he had moved the spot to the location of his choice, indicate that in some other manner, perhaps by pressing a button, or perhaps simply by lingering at the spot longer than if he were just passing by it. But Eric's occupational therapist wanted him to engage in a somewhat wider range of movements than the joystick would allow, and so still another solution was sought.

Perhaps Eric could point directly to the choices he was offered on the computer's screen. A light pen allows a user to give information to the computer by pointing directly to locations on the screen which may contain words, pictures or other information. The light pen communicates back to the computer the location (if any) on the video screen toward which it is aimed. Another way of reporting the same kind of spatial information to the computer is through a touch sensitive screen. Instead of mere pointing, direct contact must be made. Touching with a finger is a natural way of giving spatial information or choosing among visually presented options. Both light pen and touch sensitive screens are commercial devices available on many microcomputers.

A light pen could be strapped to Eric's head wand, or Eric could use the wand itself to make contact with the touch sensitive screen. In addition to providing alternatives to the keyboard, these input schemes might allow Eric to communicate a crude kind of spatial information.

One problem with these schemes is that they link communication closely with head position and therefore, to an extent, to eye position. Sometimes this choice is unavoidable, but Eric's occupational therapist thought it worthwhile to investigate ways to leave Eric's eyes maximally free to observe the environment. The idea of Eric using his hand to point at the screen was now considered. Although Eric's coordination was insufficient for typing, when he had been much younger, he had pointed adequately to a picture-board laid horizontally on his lap. Unfortunately, the vertical orientation of the screen made it virtually impossible for Eric to point accurately to items on the screen. Nevertheless, the idea of using his hand for spatial information was not abandoned. Instead, Eric was provided with a graphics tablet, a device on which one can draw or write freehand, and which then communicates the shape of that movement accurately to the computer. The information can then be displayed on the computer's screen or can be used to control some other output device, for example, a voice synthesizer. The tablet is a flat surface and can be placed at whatever angle best suits the individual. To help Eric hold the graphic tablet stylus, his occupational therapist used a quick hardening compound and some Velcro and molded a special grip to fit Eric's hand and the stylus. Eric could then use this device to drive a spot of light around on the computer screen, much as he might have done with a joystick, to indicate which menu items on the screen he wished to choose.

Some thought now had to be given to software. In order to accommodate Eric's poor hand coordination, either the choices on the screen had to be widely separated spatially—offering a poor selection at any one time and relatively slower access to all the sub-menus that expanded Eric's choices—or some program had to be written to make better use of his movements. The latter approach was taken. Since Eric was rarely able to keep his hand still while pointing to a target, the computer "averaged" the position of his hand as reported by the tablet over roughly 1.5 seconds, and took that average position as Eric's intention. This scheme was not initially faster than head-wand pointing, but allowed Eric a great deal more flexibility. An added benefit of this approach was that he learned to draw, much as Tammy did, by moving the stylus—which came to be referred to as his "iron"—around on the tablet. Incidentally, though the research toward developing this aid was initiated at a university, much of the programming for it was done by a high school student.

Inexpensive devices capable of sensing and reporting three-dimensional movement were just becoming available at around the time that Eric's communicator was developed, and today these promise even greater speed and flexibility than Eric's iron. The least expensive ways of sensing three-dimensional movement require that the user move a flexible mechanical "arm" that can sense the degree of bend at each of its joints, and thus compute the position in which its end has been placed. However, the sensors in the cheaper devices of this kind are mechanical linkages, which thus limit their usefulness in some applications. For example, Eric's erratic motor movements are likely to be hard on such linkages, and the physical range of the device is rather small. However, for another child, this technology's simplicity and low cost could make it very useful. For Eric, another technology that senses position optically would probably be more appropriate. Because it does not contain moving parts, it is much more durable and allows far wider range of movement.

Of course, there are even other ways of sensing body movement and position. For some children, feedback on their head position helps them to learn better head control. A crude, but extremely cheap, indicator of the position of an arm or the angle of inclination of one's head can be provided by a simple mercury switch. This simple device, like a carpenter's level, is sensitive to its orientation in space. It can be connected to the computer easily and quickly where the game buttons normally plug in.

Gill

Gill has a variety of perceptual and perceptual-motor problems, an attention deficit disorder, and delayed emotional development. When he was learning to write, a device was designed that allowed him to trace a path while the computer monitored his performance. Originally, using a light pen was considered, but that required Gill to do his writing and shape-tracing on the vertical surface of the computer display. The graphics tablet was another obvious choice, but was judged too expensive a piece of equipment for such a simple task as tracing a path. Instead, a very simple and cheap light-sensitive probe was developed and was plugged into a **game-paddle**

input port on the computer. With this probe, he could trace along a broad, black path drawn on a white background, and the computer could know when he strayed off the path. Popular magazines on computing often show how to build such devices and provide the principles by which others can be invented. Again, a high school student did the work; this time as part of a project for a science class.

Next, games with complex paths to follow and interesting computer responses to his success were constructed for Gill. The best part was that Gill, his friends, and their teacher could make up their own paths with nothing but a broad marking pen, and then play with those paths.

Unusual Interfaces

What does one do for an individual whose sense of control or causality is so weak and whose attention is so unfocussed as to preclude tasks that require cooperation, intention, or external orientation? There are now several technologies capable of sensing proximity. The ultrasonic devices that focused the earliest self-focussing Polaroid cameras are no longer the state of the art in self-focussing photographic equipment, but are very reasonable experimental tools for controlling computers. An older proximity-sensing technology, the **theremin,** gave us the wierd electronic music of

Fig. 7.1 Path drawn on paper for Gill to trace over.

the 1950s. Simple circuits based on this technology can also be built as inputs to the computer. The electronics behind the touch-sensitive switches often used in fancy elevators can turn any conductive object into an entirely safe touch-sensitive switch.

If movement at a distance can be sensed, then even autistic rocking can be detected and used to control an attractive computer-mediated event. Once causality and attention are established, the computer's sensitivity can be increased either to require closer proximity or even contact in order to initiate or control the event, or to make use of variations in proximity. For example, the speed or color intensity of an amusing animation might increase as one approaches the TV. Once variety in behavior has been established, then control can be increased by responding differentially to the behaviors.

What about a nearly totally paralyzed individual? A noise-operated switch was designed to allow a boy paralyzed with Reye's syndrome to control the computer. A microphone mounted on the boy's "trake" so that it stayed near his mouth listened for the clucking, "tsking," or kissing sounds that the boy made (though the circuit was not sophisticated enough to distinguish among them) and reported to the computer just as if the boy had pressed a switch.

And what can the computer do with speech? Sophisticated sound-input devices, capable of limited speech recognition, are commercially available and advertised in the various hobbyist magazines. Such devices are "taught" the sound of a number of words as they are spoken by the intended user by listening to the user pronounce each of those words several times. When "training" of the device is complete, it is ready to listen to a word spoken by the user and match it to words it has been trained to recognize. Devices of this sort have been used to allow individuals with high spinal cord injuries to control their wheelchairs by talking to them. The sophistication of these devices varies considerably with their cost. The cheapest use a strategy that always "recognizes" the word spoken to them. Thus, if such a device had been taught "red," "blue," and "yellow" and someone subsequently said "chartreuse," the device would not report the new word as unknown, but would dutifully select one of "red," "blue," or "yellow," whichever sounded, to it's electronic way of thinking, closest. Also, if someone other than the intended user spoke to it, even a word that it had been trained to recognize, it might hear the word differently because of the different vocal characteristics or pronunciations of the speakers and choose a different word as the closest fit. Devices that have enough sophistication to reject a word as unknown are more expensive. All of these devices, at present, recognize words only in isolation. One cannot speak a fluent sentence, even if it were composed entirely of words "known" by the machine and have it understand the sentence. Words must be spoken distinctly and slowly one by one. Still, even the cheapest devices are enormously useful in some contexts. Microphone sensitivity protects the intended user from having the device respond to stray sounds, and a well-designed program can let the user turn off speech detection from time to time.

Computers can also speak their responses out loud. If a limited vocabulary is acceptable, the voice can even resemble a tape recorded human being in quality. Speech input and output devices can be the basis not only of control and communication aids but of educational aids as well.

Does a deaf child need any special computer adaptations? Generally not, but many programs use a beep or other sound to indicate that some important event has happened—perhaps an error condition, or a reward, or a surprise message. Though the normal operation of the program may appear not to depend at all on hearing, missing these occasional messages can sometimes be extraordinarily confusing and frustrating. In general, there is little that need be done to adapt the software, but there are simple hardware modifications that can present the student with a visible signal to indicate that the beep has occurred, such as a little microphone and amplifier to "hear" the beep and play it to a light bulb instead of a speaker.

Nancy

Of course, not all modifications require sophisticated technology. For Nancy, the young, non-reader with no motoric handicap we met in Chapter 5, most of the adaptations had to do with helping her focus her attention. Sticker labels on the keys were used to highlight the important keys. Later, a keyboard guard was tried in order to help Nancy keep from accidentally hitting two keys at a time when she carelessly placed her finger off-center while aiming at a key. The guard had holes where the keys were, so that all the keys were recessed instead of sticking out. Nancy made far fewer typing errors, but the guard prevented other students who used the same computer from typing quickly, and so it was eventually removed. Had it not been necessary to share that computer with others who touch-typed, the guard would have been a wonderful aid to Nancy, at least while she was still learning how to handle the keyboard.

On Nancy's keyboard, a soft plastic overlay was then used to help protect the keyboard and keep it free of the stickiness that so often coats little fingers. These worked well, and were inexpensive, but had a tendency to disappear or get ripped easily. When it was determined that the unprotected keyboards remained in good condition, these overlays were abandoned except for special purposes. The purpose they continued to serve was informational rather than protective. Instead of sticking labels directly on the keys, fixing them to the different overlays allowed the highlighted configurations to be changed quickly and easily. The school learned of the existence of these overlays by talking with personnel at a computer company that donated the first box of overlays to the school.

ONE SCHOOL'S EXPERIENCE

Among other reasons that pioneers are difficult to copy is that what makes them pioneers is usually some unique feature that few other places have. When one reports that a group of high school students who wanted a million-dollar computer to play with collected the money for such a machine by hustling several nearby school systems to buy into a regional computer facility that they would operate, it sounds unbelievable. Yet, that is exactly what happened in Long Island, New York, at a time when having a computer with any power meant having a very big one. Here, however, time has been on the side of the latecomers. Even if today's first-time computer buyers do not have the resources of yesterday's pioneers, they have the advantage of better equipment at lower prices.

Somewhat more modest projects are possible today, and there are still stories of students running car washes to buy a microcomputer. One school's students, at their own initiative, made $1,000 selling snacks in the halls during exam week. Soon that thousand dollars may be able to buy an even better system than the one financed years ago for much more by the ingenuity of the Long Island high school students.

Here is a brief description of one school system that got into computers early and the lessons that can be learned from it. The Commack, Long Island school system was one of the pioneers. There, until 1977, only high school students who were enrolled in computer science courses had access to the regional time-shared computer system. In 1977, the district established a pilot project funded by Title IV-C at one of its elementary schools. Through Project CAL (Computer Accelerated Learning), elementary students and teachers began to experience computer learning together.

Initially, the goal of the project was to assess the effectiveness of computer learning at the elementary level. Teachers selected programs for students, who would go to the Resource Room in pairs. Local high school students with computer experience were bused to the elementary school to introduce students and teachers to the computer and to teach programming. As the elementary school teachers and students began exploring this new resource, their excitement quickly became apparent. Some teachers sought training in programming, and assisted in staff development sessions for other teachers on computer use in classrooms. The school system considered the experiment to be a success from the beginning. Now computer use in the district has expanded to include all the schools.

Today the district develops all its own software. High school students are paid to program learning activities, which Dolores Shanahan, the computer learning specialist, develops and supervises. The number of high school students who developed sufficient skill to perform services like these has increased as access to computers became more widespread. Because out-of-district administrators, specialists, and teachers have made so many requests and shown so much interest, the district's Board of Education has established a nonprofit corporation, COMCAL, to distribute their microcomputer software.

The microcomputer project, initiated at the elementary level, expanded into the secondary schools in 1980. At the elementary, junior high, and high school levels, the computers are housed in Learning Resource Centers, where they can be accessed by all students—gifted students, regular students, and students in remedial and special education programs.

This degree of accessibility creates an intensely interactive environment: children from the different educational programs work and learn together at the terminals. Some of the children use teacher-selected software; others develop and implement language arts activities and mathematics games.

Sara, a Hearing Impaired Student

Commack staff and volunteer high school students initiated an experimental program with the Mill Neck Manor School for the Deaf (see Promising Practices in the Resources Section) during the 1979/1980 school year. The high school students learned some signing, were bused to the Mill Neck School, and served as computer mentors to the Mill Neck students. The district provided PET microcomputers and COMCAL programs. The high school students, under the Mill Neck staff's direction, created new programs for the deaf students.

Sara was twelve when she began attending a seventh-grade class in junior high. Mainstreamed for five half-days in the regular classroom, she was accompanied everywhere by an interpreter who signed her lessons to her. In the afternoons Sara attended Mill Neck. Although she initially used the computer during mathematics classes at the junior high, Sara quickly became proficient with the word processor and its multiple options. She was so facile that she became a facilitator and mentor, and, along with the high school students, introduced other Mill Neck students to the computer.

Laura, a Physically Handicapped Student

The Commack School District implemented a computer project with the physically handicapped in 1980. Students from the BOCES III physically handicapped program were selected and bused to the Commack Computer Lab, located at the Indian Hollow Elementary School. Students in the project used mathematics programs, language arts programs, simulations, and logic games. The students used the word processing and mail programs to communicate with others—handicapped and non-handicapped—in the school. Here, too, students who learned to program helped teach other students to program.

Laura was one of the students originally bused to the computer lab. Accessing the computer by means of the joystick with which she also controlled her wheelchair, Laura quickly became a proficient programmer and began introducing the computer to her classmates. Her success in teaching won her the nickname "Computer Wizard" and she became Dr. Shanahan's protege in designing individualized programs for other students.

Lessons from the Experience

When their access is not limited, interested students can develop considerable skill at computer programming. Commack and other school districts have found ways of tapping this local talent in ways that are beneficial both to the students and to the school system.

The involvement of high school students with the elementary schools serves potentially in several ways. The school system benefits both from the talent that the high school students bring, and from the clear message—to students and staff, alike—that the students are meaningfully involved, indeed essential, to the running of the schools. Commack's idea of busing high school students to the elementary school to help teach, again provided not only the content, but a message. The role-reversal—students teaching—had to be noticed. There is also a message, not just computer content, in the teaching done by the "handicapped" students. And the benefits of peer teaching, for both student and teacher, are well known. Teaching teaches. It is not the computer that has created the best features of this educational program, but the thoughtful social view of the community. The computers merely provide the excuse, the activity around which this responsible social involvement was planned.

The image of an enterprising community is complemented by the Board's establishment of a nonprofit corporation to handle the district's creativity.

However, before adopting any of the Commack and other pioneers' techniques—even the attractive and successful ones—it is important to consider the alternatives and the advantages of those alternatives. These pioneers may not have thought of or have been able to implement some of the ideas presented in this book. Technology has changed.

One aspect of the Commack story illustrates the complexity of some of the decisions that must be made. Centralized computing that all students go to has advantages and disadvantages. Bringing a variety of students together in one location may be an advantage. However, disconnecting one kind of learning from another by a walk down the corridor or a bus ride across town may be a disadvantage. Ideally, as with the private and cooperative centers in a classroom, one would have both experiences available, an option that will be very realistic as computers become more common and less expensive.

EPILOGUE

Are there no failure stories? Do computers work for all students? Do all schools adopt computers as easily as Commack? The answer is "No" to all three questions.

Computers are not magic. Education's current excitement with them could conceivably be as short-lived as the excitement over teaching machines or the "new math" was in the 1960s. But then, it is precisely because the computer's frequent successes are not at all magical, but the understandable results of augmentation of one's capabilities, that we believe it will last as a tool to enhance education. For some children, in some circumstances, the computer can be wonderful. In other circumstances, it may be of negligible value, and in still other circumstances, a negative factor altogether. Decisions about what computer experience is good at what time and for whom are like all other educational decisions: they rest entirely on human sensitivity to the students and good judgment by good teachers. None of this is easily prescribed in a book.

Whether a school district or individual school is able to obtain computers and train teachers in their use also depends on factors that cannot always be arranged. For example, it may require:

- one interested teacher who obtained a small grant, or
- a parent who donated a computer or persuaded an employer to donate one, or
- proximity to a computer company interested in community support or public relations to assist local schools, or
- a local university which needed a school site for a computer education project.

Success or failure of an existing school program using computers may depend on:

- ability and inclination of the school board to include funding for hardware, software, and staff in its budget
- availability and willingness of knowledgeable teachers or parents to spend time helping others get started
- overcoming initial fear and anxiety about using the computer and about the teacher's role in relation to it
- accessibility of the computer and appropriate computer-based activities to all students.

For many students with special needs, there are other factors that determine whether or not they obtain access to appropriate educational technology: chance and public policy.

Two stories in which only the names have been changed illustrate the chance factor well. Fred has spastic quadriplegia. He was 17 when a group from a large university approached his school—a private school for physically handicapped students—and began working with him using the Logo computer language. Severely limited in his ability to speak or write, Fred found the computer to be a means of communication and a tool for learning such as had not been previously available to him. A nationally aired television program featured Fred's story and donated a computer to the school. Eventually, a computer company donated a computer to Fred for home use. By contrast, Marcia became paralyzed in an automobile accident at the age of 18. A friend of hers, David, worked in the computer field and helped her make contact with Alan, the head of a division of a large computer company. Alan personally supported Marcia's cause and persuaded the company to give her a home computer. However, before the donation occurred, Alan was transferred and the entire transaction was lost in the shuffle.

While individual students or schools benefit from fortuitous encounters with power, concern, or wealth, it is a precarious existence, depending on whim, generosity, and publicity. The vast majority of students are not the fortunate recipients. That brings up questions of public policy. While medical insurance policies standardly cover such technologies as eyeglasses, wheelchairs, and hearing aids, it remains extremely difficult to fund expressive communication aids and nearly impossible to fund educational aids, though all these interventions may have indistinguishably similar impact on the life of an individual.

It is too expensive in the waste of human potential—and even in the dollar cost to society—to leave it to chance whether an individual will receive an appropriate education. Current levels of funding from public and private sectors as well as standard medical insurance do not provide what is needed.

While the computer is not a necessary item in everyone's life, it can, for some, be a tool for communication, recreation, exploration, learning, and vocational access. Where it is serving a necessary function, it can no more reasonably be denied to someone who needs it than entrance to a library can be denied for lack of a wheelchair ramp.

The goal of education is a productive and estimable life. If a thousand dollars worth of computer equipment can make that possible, it is very near-sighted not to make the investment. But such a productive and estimable life is not and will not be the result of any computer. After all, the computer is only a tool, and it can be used to damage as well as to aid. If a person has an interesting and satisfying life, it will be the result of caring family or friends, and good learning and work environments. The life is the activity, not the equipment.

What that means for teachers is that neither their jobs nor their dreams of a full life for all their students has changed. What has changed is the chance of achieving those dreams.

Glossary

Amyotrophic lateral sclerosis A degenerative neurological disease causing progressive weakness and paralysis (but no impairment of mental function), and usually ending in death over the course of two to three years. Abbreviated ALS, sometimes known as Lou Gehrig's disease.

Arthrogryposis A congenital, nondegenerative disease in which children are born with weak muscles compounded by stiff joints and often with other congenital anomalies, but no impairment of intelligence or speech.

Artificial intelligence The branch of computer science from which Logo grew. The name of this science derives from its attempts to simulate, using machines, the behavior that is regarded as intelligent in people or animals. If this was ever the only goal of the field, it no longer is. Cognitive psychology and artificial intelligence have grown very close, and together are sometimes referred to as cognitive science. They study such disparate processes as language understanding, visual perception, knowledge acquisition, reasoning, and self-correction.

Athetosis A movement disorder which is characterized by involuntary and continuous, slow, sinuous, writhing movements, often especially severe in the hands. This is a common subtype of cerebral palsy and may be mixed with varying degrees of spasticity.

Autism A marked behavioral disturbance characterized by apparent inability to relate normally to people, poor (or absent) communication, rigid preoccupations with sameness and, sometimes, apparently total insensitivity to sights and sounds (as if blind and deaf) or even to pain. Once theorized to be caused by inappropriate parenting, autism is now generally agreed to be the result of a severe and early disorder in the development of language and extralinguistic forms of communication. This communication impairment may be recognizable in infancy even before the development of speech in normal children and is often compounded by other perceptual and cognitive handicaps that impair normal relations with people, and by the emotional overlay that naturally accompanies the distorted and chaotic world that such a communication failure would produce.

Blindness A generic term for severe visual impairments, but not restricted in meaning to total absence of sight. Legal definitions of blindness in many places include persons who have, after correction, a visual acuity of 20/200 or less in the better eye, and a visual field limited to about a quarter or less of the normal. Thus, some blind people can read print, if the print is large enough and close enough. The word "blindness" is also used to refer to a variety of impairments of visual perception that are not the result of limited acuity or field, e.g., word blindness and other neurological conditions that prevent the individual from recognizing some aspect of a particular stimulus, though otherwise apparently able to "see" perfectly.

Bliss symbols An ideographic writing system used as an aid to communication for some individuals who it is thought cannot use print as effectively. Individual symbols have a meaning by themselves (e.g., a heart, by itself, stands for the concept "emotion") and can combine with other symbols to convey variants on the central meaning (e.g., happy, sad, want, like, or love).

Cerebral palsy A nonprogressive disorder of movement resulting from damage to the brain occurring before the age of three years. One of the most common causes of CP is severe oxygen deprivation (hypoxia) at or around birth (perinatally), due, for example, to Rh incompatibility or prolapse of the umbilical cord secondary to traumatic presentation at birth. Other causes include prenatal rubella, and head trauma or other injury to the brain before the age of three. Cerebral palsy is not necessarily accompanied by any cognitive defect, and recent research suggests that mental retardation may be no more expectable among those children whose cerebral palsy is due to perinatal hypoxia than in the population at large.

Computer mail   See Electronic mail.

Deafness   Partial or complete lack or loss of hearing. Some people prefer the term "hearing impairment," as it does not imply totality and finality. Other people prefer the term "deafness" precisely because of its frankness in not denying the psychological reality for many deaf individuals. Like blindness, the term is used with a variety of meanings, but, in general, is applied when hearing is insufficient, even with amplification, for the accurate and reliable perception of speech. Most deaf children have hearing parents. A still too common cause of deafness is prenatal rubella (german measles). See also Prelingual deafness.

Disability   One of a large number of terms used to refer to or euphemize people who are, in some way, different from, and often seen or treated as inferior to, the norm. The very number of these terms—disabled, special, handicapped, exceptional, etc.—attests to society's discomfort with the category, and the periodic shift from one term to another suggests an attempt to find a non-obnoxious way to refer to an anxiety-provoking difference. Some people prefer "handicap" because it does not suggest finality, merely an impediment, like a golfing handicap which may give the competitor the edge, but not necessarily the game. Other people prefer "disability." "Special needs" and "exceptional" are more inclusive (we all have needs) but also more euphemistically bland.

Disk   A disk-shaped electronic storage medium for storing computer programs or other information magnetically much as an audio tape stores recorded music.

Disk drive   A peripheral device that can record programs or other information on disks and can retrieve that information for later use.

Electronic mail   A computer-based electronic message system that enables one person to compose a message, address it to another person, and ship it electronically to that person's computer "mailbox." The message waits in the recipient's mailbox until that person requests his or her computer to check that mailbox for any received mail Electronic message systems are in very wide use in larger offices and businesses, in government, and in universities. Non-corporate private message systems and even some public systems are beginning to arise, and will increase rapidly.

Game paddle An input device that resembles a dial which one may turn left or right in order to control a game or some other behavior of the computer. The game paddle may be a useful alternative for typing as a way of communicating with the computer.

Game paddle port The word "port" is used to refer to an external electronic connection to the computer through which information may be sent to and from the computer. A game paddle port is the connection into which a game paddle can be plugged. Often other devices, suitably designed, can be plugged in to this ready-made port.

Graphics tablet A flat drawing tablet capable of sensing the position of a finger, pencil, or special stylus on its surface, and communicating that two-dimensional spatial information to the computer.

Handicap See Disability.

Hearing impairment See Deafness.

Impairment Literally, a deficiency or inadequacy compared to a standard. Psychologically, the word has many meanings and is sometimes used in effect to minimize the severity of some condition (see Deafness, for an example). The term also contains an implicit external standard against which the "impaired" person is measured and fails, and may therefore be perceived as evaluative or judgmental. As with any term for a disability the acceptability of the term varies from group to group and over time.

Joy stick An input device somewhat resembling an automobile's manual gearshift lever with which one may indicate directional information, sometimes including even the magnitude of a directional vector. The joy stick is often a useful alternative to typing as a way of communicating with the computer, not only for games, but for any task in which spatial layout of information is important.

Keyboard The universal input device on computers, often arranged much like the keyboard of the typewriter, but generally containing extra keys that serve special purposes related to the computer. Though the keyboard is the most common device through which a person's intentions are communicated to the machine, it is not the only one, and often not the best one. See Graphics tablet, Joy stick, Light pen.

Learning disability A term referring to any of a large class of disorders affecting learning, some of which are quite specific in their form and appearance, but many others of which are defined more by what they are not (e.g., a reading failure that is not due to general retardation, lack of knowledge of the language, or visual impairment) than by what they are. The definitions of many learning disabilities and the terminology by which they are classified—even the existence of specific disabilities—are in a state of rapid change with terms as different as minimal brain disfunction (MBD), developmental lag, perceptual impairment, and hyperactivity used as ways of classifying the same behavioral observations. The more common learning disabilities are fairly specific to reading (sometimes referred to as dyslexia) and attention (referred to variously as attentional deficit, hyperactivity, and the rather diffuse "behavior disorder"). Dysgraphia and dyscalculia are names sometimes used for two less common learning disabilities. The first refers to difficulties with writing (either handwriting itself or the whole area of expressive communication through print) that do not adversely affect reading. The second refers to difficulties with arithmetic and is commonly associated with certain very specific but relatively inconsequential coordination problems.

Light pen An input device that allows a computer to sense where, on its video display, the device is aimed, and with which one may indicate positional information. The light pen is often useful for selecting among items presented on the display.

Mental retardation Generalized subnormal intellectual development as evidenced by impairments in learning, social adjustment, or both. Mental retardation, though classified as a developmental disorder, can be of functional origin, the latter resulting from abnormally restricted experience and mental stimulation, or from other early and chronic mentally depressing causes. Many severe handicaps, especially when communication is significantly impaired, are misdiagnosed as being compounded by mental retardation. Even with some syndromes classically associated with mental retardation, e.g., Down's Syndrome (mongolism), it has begun to appear that not all who are affected by the syndrome are retarded, and that some of the retardation that does appear is secondary to the experience, treatment, or testing of the individual and not a primary attribute of the condition.

Motoric impairment A disorder of movement.

Music synthesizer An electronic device capable of generating musical tones. Sophisticated synthesizers can mimic the sounds of specific musical instruments or combinations of instruments and can generate tone qualities not found in conventional instruments.

Network An electronic communication system linking computers (often large numbers of them) to each other and enabling each to transfer information to and from each of the others. Electronic message systems (see Electronic mail) often rely on computer networks for any message transfers to remote locations.

Prelingual deafness Deafness occurring before a time when spoken language is well developed (generally taken to mean before the age of three). As prelingual deafness sharply reduces the amount of contact a child has with the surrounding spoken language, it is frequently associated with moderate to profound weaknesses in all aspects of that language, including reading and writing. According to one statistic, roughly a fourth of deaf children under the age of 18 are prelingually deaf.

Proximity sensor A device capable of sensing nearness or approach and potentially useful as an input device that does not require contact or even attention.

Processor The "brain" of the computer that controls the interpretation and execution of stored instructions.

Prosthesis A device to replace the function of an impaired or missing body part. Artificial arms and legs are prostheses for manipulation and walking. By analogy, computers can be informational prostheses, helping people manipulate information or ideas that their ability with their unaided fingers, eyes, or ears might not let them perform—a reading prosthesis, or drawing prosthesis, for example.

Robot A mechanical device that operates automatically or under computer control and that performs tasks modelled after human or animal behavior. Robotic devices have been developed for manipulating dangerous materials, working in dangerous environments, or performing tasks too delicate or tedious to be practical for people. Applications of robots as prosthetic devices are developing.

Spastic Spasticity is a state of abnormally increased tone or tension in a muscle, and is a common subtype of cerebral palsy. The popular American use of "spastic" as a jocular equivalent of "clumsy" conjures up an image of jerky or wild movements, whereas true spasticity tends to be characterized by poorly controlled but very slow and limited movements. In England, "spastic" is sometimes used to refer to cerebral palsy in general.

Spina bifida A congenital defect in the bony encasement of the spinal cord, often with protrusion of the meninges through the cleft, and generally resulting in partial or total paralysis below the site of the defect. Mental retardation is sometimes but not always associated with spina bifida.

Speech synthesizer A device capable of generating the phonemes (elementary speech sounds) of a language and combining them intelligibly to simulate speech. Many speech synthesizers are capable of accepting their text input spelled in the usual way, and contain within them a large number of pronuciation rules to translate from that spelling (traditional orthography) to the appropriate set of phonemes to pronounce (not always possible without understanding the context within which the word appears). These are sometimes called text-to-speech devices. Other synthesizers require that the pronunciation of the word be given explicitly, phoneme by phoneme, with special phonetic spellings. In either case, since the speech is synthesized electronically and not merely recorded (see Speech digitizer), any text may be spoken by the device as long as it is in the language (e.g., English) for which the device was designed. Also, because the speech is entirely synthetic, overall speech quality depends on the sophistication of the particular device, but all synthesizers to date have a "machine-like" quality to their voice.

Speech digitizer A device capable of recording pronounceable chunks of speech (e.g., words, phrases, and, rarely, syllables) in a way that can be stored in computer memory and later reproduced in any arbitrarily selected order. Since the speech is actually recorded, the speech quality can be very high, resembling the voice of the original speaker quite accurately, depending only on the fidelity of the particular digitizer used. However, since no synthesis is involved, the vocabulary is ultimately limited by what has been recorded, which, in turn, is a function of available memory, and consequently, cost. Attempts to increase vocabulary without taxing memory, by selecting an optimal set of small, highly combinable word-parts (e.g., syllables) from which to construct a large vocabulary have had some limited success, but the intelligibility of words constructed from concatenated parts can be very poor.

Speech recognition device A device capable of accepting words spoken into a microphone as input, and identifying those words by comparing them to a limited vocabulary already encoded in its memory. Because it can identify its input (what is said to it), its output can be used to control a computer. The size of the vocabulary that can be recognized and the accuracy of recognition depend both on memory size and processing sophistication of the device, and therefore depend heavily on cost. Recognition of continuous speech (words spoken naturally and not separated by pauses), especially with more than a very small vocabulary, is quite difficult. Only a few such systems as yet exist; they are on very large computers, and even they are limited both in vocabulary size and length of phrase that can be recognized. Many inexpensive devices, however, are capable of recognizing a few dozen words spoken individually. Since the small vocabulary is selectable by and tailored to the user, these devices can be used even by people with defective but consistent speech, and a sufficient number of words is generally available for representing commands to the computer, letters of the alphabet, etc.

Theremin   Possibly the earliest of music synthesizers, having been invented in the 1920s. The pitch and volume of the tones produced by this device were controlled by the proximity of the player's hands to two metal antennas. See Proximity sensor.

Touch sensor   An input device that recognizes contact, without necessarily requiring pressure or movement. See also Touch sensitive screen.

Touch sensitive screen   A surface or overlay for a video display screen that recognizes and communicates to the computer the location at which it is touched. When information (e.g., letters, numbers, menu choices) is displayed in different locations on a touch sensitive video screen, the user can select among the items presented, thus communicating with the computer, merely by touching the different locations.

Visual impairment   See Blindness.

Voice recognition device   See Speech recognition device.

Resources

INTRODUCTION

This Resources Section provides myriad sources of people and places involved in computer applications for children with special needs. Yet, technology is moving so swiftly that new sources are springing up on an almost daily basis. Thus, the resources that follow by no means constitute a comprehensive listing, but rather represent a sampling of what is available.

Moreover, the reader will notice that the majority of resources we've selected deal with severe sensory and motor handicaps; fewer would seem to apply to the learning disabilities and behavioral disorders that are most common in schools. This, in part, reflects our own emphasis on communication and access problems, which for severely disabled students must be addressed before any other educational intervention can be effective. In part, this also reflects our assessment of the relative need for various resources; we have included primarily those we believed would be less familiar to most teachers and/or most helpful in locating still other resources.

We have, of course, also included some resources specific to the special needs most often found in schools: dyslexia, hyperactivity, and a range of learning disabilities. Most of these can be used to find other resources applicable in these situations. Moreover, it often happens that tools or strategies developed for a particular special need turn out to be very appropriate for another. As examples, speech output devices developed for the blind may be very useful for students with reading disabilities, "English as a second language (ESL)" materials may be quite valuable for deaf or language disabled students, and hardware adaptations for motorically impaired students may work quite well with very young children who experience difficulty with a computer keyboard for any number of reasons. Thus, we think you will find many ideas and resources for addressing the full range of special needs in educational settings.

We encourage you to contact the resources listed here to take advantage of the information they have to offer. We hope this information will inspire you to become involved in this exciting new era of education and special needs.

For those who want a broader range of resources than is offered here, see the other books in Addison-Wesley's *Computers in Education* series, including: *Practical Guide to Computers in Education*, *Computers and Reading Instruction*, and *Computers in Teaching Mathematics*. These are available from Addison-Wesley Publishing Company, Reading, MA 01867, 617-944-3700.

Special thanks to Helen Pollack, whose responsibility it was to compile these resources. Additional thanks to Marilyn Martin and Carol Nuccio for their input.

RESOURCES CONTENTS

ASSOCIATIONS AND ORGANIZATIONS	208
TRAINING AND RESOURCE CENTERS	212
PROMISING PRACTICES	215
PUBLICATIONS	
Periodicals	223
Special Issues of Magazines	227
Reports and Articles	228
Bibliographies	232
Bibliography of References in This Book	233
Books on Computers, Education, and Special Needs	237
Books on Particular Disabilities	239
ON LINE NETWORKS AND DATABASES	240
SOFTWARE	
Selected Software for Special Needs	243
Software Mentioned in This Book	246
Software Directories	250
HARDWARE	
Output Devices	251
Input Devices	255

ASSOCIATIONS AND ORGANIZATIONS

Membership in an organization or special interest group is a way of keeping up-to-date on new developments in technology as it relates to special needs. Organizational publications offer results of recent research, list workshops, and offer practical hands-on information. While there are many organizations that address various special needs populations, we have included the following groups because at the time of this writing they are specifically involved in research, development, or the dissemination of information regarding computer technology and special needs.

For the names and addresses of the myriad computing and educational computing organizations, see the Resources Section of Addison-Wesley's *Practical Guide to Computers in Education*.

American Association on Mental Deficiency (AAMD)
5101 Wisconsin Avenue, N.W.
Washington, DC 20016
(800) 424-3688

This professional association is for those working in the area of mental retardation. Members include psychologists, physicians, social workers, as well as educators. Currently, the association is involved in promoting the use of and providing information regarding computers and the mentally retarded. AAMD publishes two professional journals, *The American Journal on Mental Deficiency*, and *Mental Retardation*. Dues are $60 per year.

American Federation for the Blind (AFB)
Technological Development Department
15 W. 16th Street
New York, NY 10011
(212) 620-2080

The AFB serves the needs of the blind and visually impaired through rehabilitation services, legislative consultation to government agencies, and advisory services to schools and local agencies. The AFB is a good resource for information on computerized speech output devices.

American Speech-Language-Hearing Foundation
10801 Rockville Pike
Rockville, MD 20852

ASLHF is committed to the improvement of knowledge related to: prevention of communication disorders; normal and disordered communication; clinical service delivery; and professional development. The Foundation seeks to enrich and expand the bases on which the profession is built—through support of research, recognition of outstanding clinical achievement, and programs of professional development, including those dealing with technology in speech-language pathology and audiology.

Association for Special Education Technology (ASET)
Attention: Mary Ventura
P.O. Box 152
Allen, TX 75002

ASET is a national affiliate of the Association for Educational Communications and Technology. Its purpose is to bring together various disciplines that are concerned with improving the use of technology in special education. Among its activities are the promotion of federal legislation for technology in special education, the development of new technologies, the adaptation of materials, and the identification of instructional needs of special education. The organization's publications include the *ASET Report* and the *Journal of Special Education Technology* (see **Publications: Periodicals**).

The Committee on Personal Computers and the Handicapped (COPH-2)
2030 Irving Park Road
Chicago, IL 60618
(312) 477-1813

This organization serves to locate, evaluate, and share information on products available to handicapped persons. Members receive the *COPH Bulletin* (see **Publications: Periodicals**), the newsletter *Link and Go*, and access to the *COPH-2 Resource Box*. The annual membership fee is $8.

Computer Users in Speech and Hearing (CUSH)
Dr. James L. Fitch
Department of Speech Pathology and Audiology
University of South Alabama
Mobile, AL 36688
(205) 460-6327

A spinoff of the American Speech and Hearing Association's Committee on Technology, CUSH provides professionals in the field with information on computer use with individuals who have communication handicaps. CUSH publishes a newsletter quarterly and a software registry twice each year, which list software that has been both privately and commercially developed. Membership is $5.00 a year.

The Council for Exceptional Children (CEC)
Attention: Lynn Smarte
1920 Association Drive
Reston, VA 22091
(703) 620-3660

The Council provides interested educators and parents with information about the use of microcomputers in the advancement of handicapped and gifted children. CEC is organized at the local, state, and national level, and has twelve special interest divisions in which members may participate. CEC members receive three periodicals, *Exceptional Children*,

Teaching Exceptional Children, and *UPDATE* (see **Publications: Periodicals**). Members are also entitled to discounts on CEC publications, computer searches, and products included in the CEC Information Center. Dues vary by state and membership status.

Education TURNKEY Systems, Inc.
Charles L. Blaschke, President
256 North Washington Street
Falls Church, VA 22046-4544
(703) 536-2310

TURNKEY provides local education agencies with information about technologies, trends, and potential applications for special education. It is also concerned with the marketing of educational software and works to ensure that the products available are beneficial for the handicapped learner. MEAN, a division of TURNKEY, operates the Computer Bulletin Board on SpecialNet (see **On Line Networks and Databases**).

EduTech
Susan Elting, Director
Log AB
JWK International Corporation
7617 Little River Turnpike
Annandale, VA 22003
(703) 750-0500

EduTech is intended to encourage interaction between technologists and educators to better meet the needs of special education, and to keep educators informed about issues, individuals, agencies, and organizations through direct mail, publications, and an electronic bulletin board on the SpecialNet system (see **On Line Networks and Databases**).

International Council for Computers in Education (ICCE)
135 Education
University of Oregon
Eugene, OR 97403
(503) 686-4414

ICCE is a nonprofit organization for people interested in instructional computing at the pre-college level. It publishes *The Computing Teacher* and numerous booklets which frequently address special education applications (see **Publications: Special Issues of Magazines**).

Minnesota Educational Computing Consortium (MECC)
Attention: Karen Jostad
3490 Lexington Avenue N.
St. Paul, MN 55113
(612) 638-0600

MECC, which produces and provides software for students, teachers, and administrators, has seven courseware packages available for the physically handicapped. These include: *Special Needs Vol. I—Spelling* (a spelling practice package) and *Special Needs Vol. II* (a package with drills and simulations in several disciplines for motor impaired students). There are also four packages in Bliss Symbolics for nonvocal students. MECC has produced *Guessing and Thinking*, a courseware package for hearing impaired students.

National Association for Visually Handicapped (NAVH)
Lorraine Marchi, Executive Director
305 E. 24th Street
New York, NY 10010
(212) 889-3141

NAVH offers information and referral services to children and adults who are partially sighted. NAVH also participates, in cooperation with commercial manufacturers, in the field testing of new technological optical aids.

Sensory Aids Foundation (SAF)
Attention: Susan H. Phillips
399 Sherman Avenue, Suite 12
Palo Alto, CA 94306
(415) 329-0430

SAF is a nonprofit organization that sponsors research projects in technology and the handicapped. In addition, SAF publishes a quarterly newsletter, *Sensory Aids Technology Update* (see **Publications: Periodicals**), offers consultation services, and identifies employment opportunities for the handicapped.

Telecommunications for the Deaf, Inc.
Attention: Barry Strassler
National Association of the Deaf
814 Thayer Avenue
Silver Spring, MD 20910
(301) 589-3006

Organized as a nonprofit organization in 1968, TDI works to develop compatible standards for telecommunication devices for the deaf and works with government agencies and legislative bodies to encourage use and support of such devices. Recent efforts have focused on bridging the gap between established telecommunications for the deaf (devices for the deaf to send and receive typed messages via the telephone and telephone lines) and microcomputers, which can provide further opportunities for communication at home and can enhance the work environment for the deaf.

TRAINING AND RESOURCE CENTERS

The following groups and places provide workshops and other types of assistance regarding computers and special needs. For general computer workshops and resources, see the Resources Section of Addison-Wesley's *Practical Guide to Computers in Education*.

Castle Priory College
Mrs. Joyce Knowles, Principal
Wallingford
Oxford
United Kingdom
0491-37551

Castle Priory College opened in 1965 and is designed to offer study for professional and voluntary workers involved in the care, education, and treatment of the handicapped. It has been the site of workshops on the applications of microcomputers for the handicapped and has offered courses on technology and disabled children.

Closing the Gap Workshops
Dolores Hagen, Director
P.O. Box 68
Henderson, MN 56044
(612) 665-6573

These workshops are custom-designed to meet the needs of special education professionals to receive hands-on experience to learn how to use microcomputers effectively.

Decision Support Systems Ltd.
Attention: John J. Willson
301 Kingsway Garden Mall
Edmonton, Alberta, CA T5G 3A6
(403) 471-5471

Currently underway is a project to catalog various information sources, suppliers, courseware, and courseware authors who provide software written or adapted for the handicapped.

Eastern Pennsylvania Regional Resources Center for Special Education
1013 W. Ninth Avenue
King of Prussia, PA 19406
(215) 265-7321

Eastern region educators involved with special education may use the center for research in educational technology, teacher training, and software review. The center includes a library and demonstration lab and offers information retrieval searches on major online databases.

Educational Computing Group
Laboratory of Computer Science
MIT
545 Technology Square
Cambridge, MA 02139

Offered here is information on Logo and a videotape series on using Logo with special needs children.

Laureate Learning Systems, Inc.
Mary Wilson, President
1 Mill Street
Burlington, VT 05401

This company develops software and provides services and training to support innovative uses of microcomputers in special education.

LINC Resources, Inc.
Jeanine Tolbert or Emmett Crawley
1875 Morse Road, Suite 225
Columbus, OH 43229
(614) 263-5462

LINC provides technical assistance and marketing services for special educators who are developing materials under grant or contract from the U.S. Department of Education. Workshops are held each spring dealing with marketing factors that relate to product development as well as to the commercial publications potential of specific products.

Microcomputer Information Coordination Center
Judy Wilson, Project Coordinator
University of Kansas Medical Center
Room 139 CRU
39th and Rainbow Boulevard
Kansas City, KS 66103
(913) 588-5985

M.I.C.C. is designed to facilitate telecommunications and microcomputer applications in special education throughout Kansas. *SpecialNet* is accessible at 88 special education sites, and includes five statewide bulletin boards. The project also coordinates a statewide software evaluation system. This project is funded by the Kansas State Department of Education.

The Mill Neck Foundation, Inc.
℅ Mr. Louis W. Frillmann
P.O. Box 100
Mill Neck, NY 11765
(516) 922-3880 (Voice/TDD)

The Foundation (located on the campus of the Mill Neck Manor Lutheran School for the Deaf) sponsors a Mobile Telecommunication Van that serves as a Training and Resource Unit. The van is equipped with an audiometric testing booth, all the latest telectronic devices for the deaf(TDD), decoders for TV captions, light and signal systems, and a microcomputer to demonstrate the use of computers in deaf education and for networking. The van travels extensively in Metro New York offering hearing screenings, awareness training, and demonstrations, and goes on the road for special events.

National RETOOL Center
Attention: Josephine Barresi
The Council for Exceptional Children
1920 Association Drive
Reston, VA 22091
(703) 620-3660

The teacher education division of CEC, the National RETOOL Center promotes continuing education through teacher training sessions conducted nationwide. They sponsor a national conference on Technology in Special Education: Management and Instruction, as well as a national software competition for commercial and individual developers.

Technical Education Research Center (TERC)
Attention: June Foster
44 Brattle Street
Cambridge, MA 02138
(617) 547-0430

TERC conducts research and development projects to determine the most effective application of microcomputers to special education. They also develop software and hardware adaptations appropriate for special needs learners as well as conduct workshops for professionals involved in special education.

TRACE, Research and Development Center for Severely Communicatively Handicapped
Attention: Chris Thompson
314 Waisman
1500 Highland Avenue
Madison, WI 53706
(608) 262-6966

TRACE offers information on communication, writing systems, and access to computers for severely physically handicapped individuals. Gregg Vanderheiden, Director, presents workshops throughout the world on applications and techniques to improve communication and interaction for non-vocal children and adults. The Center also publishes two resource books, *The Non-Vocal Communication Resource Book* and *The Rehabilitation Aids Resource Book*.

Utah State University
Attention: Kim Allard or Bob Reid
Logan, UT 84322
(801) 750-2032

Utah State University has developed a database of references related to computers in special education and a database for reviews of microcomputer software for special education teachers.

PROMISING PRACTICES

Technology has probably done more to integrate learners of great diversity and to provide equal opportunity for education and jobs than most legislation in the last twenty years. The space race of the sixties introduced the term *microminiaturization*, and with it came the dawn of microcomputers and the promise of a new world of communications. Science moved out of the laboratory and into the classroom, the workplace, and the home. Hobbyists fiddled with surplus government equipment and eventually produced a new communications device for the deaf by connecting the teletypewriter and the computer to create electronic bulletin boards and message systems; a mechanical turtle moved around the floor and helped autistic children relate to their environment; the magic of a new language called Logo helped students with cerebral palsy write with clarity and purpose, even to learn to program and show their untapped knowledge of mathematics.

The following section contains descriptions of some of the promises of technology, and introduces people and places who may be helpful resources in the reader's own quest. That these practices are limited by required funding, grants, and gifts is evident. We hope the next twenty years will see a reduction in the economic barriers and grant the opportunity of access to technology to all who desire or need it.

The Alternative Communication System Project
Washington Research Foundation
Suite 322
U District Building
1107 N.E. 45th Street
Seattle, WA 98105

Special input software for children with physical disabilities developed in the Maplewood Apple II Computer Project, Edmonds, Washington, and in the Alternative Communication

System Project, University of Washington is available for the Apple II and IIe computers. Disks include *Motor Training Games*, *Scanning and Morse Code Practice Programs*, *Academics with Scanning*, and a *Special Inputs* package.

Amateur Radio Research and Development Corporation (AMRAD)
Attention: Howard Cunningham
P.O. Drawer #6148
McLean, VA 22106

Federal funding has enabled AMRAD, a nonprofit organization, to develop interfaces that will allow microcomputers to replace teletype terminals as communication devices for the deaf. Their first interface was available for the Apple.

Assistive Device Center
Attention: Albert M. Cook, Ph.D.
California State University
School of Engineering
6000 J Street
Sacramento, CA 95819
(916) 454-6422

The Assistive Device Center provides communication, educational access, and engineering services to the physically handicapped. Following an in-depth assessment of client needs, devices for communication, mobility, self-help, and education are recommended and/or provided.

Bioengineering Program
National Association of Retarded Citizens (NARC)
Dr. Al Cavalier, Director
2501 Avenue J
Arlington, TX 76011

The research division of the NARC has undertaken a project to develop and demonstrate the use of technology enabling greater independence for the mentally retarded. With funding provided by major aerospace and computer corporations, the project contains the following components: identification of technological needs, designing and construction of devices, and application of devices.

California School for the Deaf
Attention: Dan Castle, Lab II
3044 Horace Street
Riverside, CA 92506
(714) 683-8140

The California School for the Deaf has two computer labs to assist students from 6 to 20 years of age with mathematics and language arts. The Interactive Videodisc Project combines computer and videodisc technologies to permit individual interaction between the student and a televised language program. Computer printouts show student progress and allow evaluation of language skills. Students also learn programming and have access to networking systems for information and communication.

Center for Educational Research
Attention: Richard P. Swenson, Chrys L. Anderson
44 N Last Chance Gulch
Helena, MT 59601
(406) 443-7796; (406) 443-3600

CER staff have a grant to modify computer hardware and software for use by individuals with developmental disabilities in instructional settings. They have also developed tutorial and simulation programs in budgeting and nutrition that are of special help for individuals with learning problems.

Computer Modifications for the Handicapped
Attention: Laurie S. Satre
West Central Educational Cooperative
Service Unit
120 South Vine Street
Fergus Falls, MN 56537
(218) 739-3273

The goal of this project is to provide special education students with the benefits of educational microcomputer programs. The first phase is modifying microcomputers to react to speech for the physically handicapped, to present high resolution graphics and text for the visually impaired, and to provide special earphones for the hearing impaired.

The Cotting School for Handicapped Children
Michael Talbott, Assistant Principal
241 St. Botolph Street
Boston, MA 02115
(617) 536-9632

The Cotting School, a private nonprofit day school, has been using microcomputers since 1979 to teach mathematics and to help students acquire new skills which would aid them in gaining employment. Through the use of the Logo language and selected software, students with severe physical handicaps have learned to interact with computers to solve problems and achieve a sense of control over their environment.

DEAFNET
Attention: Teresa Middleton
SRI International
333 Ravenswood Avenue
Menlo Park, CA 94025
(415) 326-6200

This two-year project is involved in helping local communities in twenty major cities with technical assistance in developing their own computerized communications network for deaf community members. Deafnet provides the software free of charge to any interested person. It may be used with any regular ASCII computer terminal or Baudot TTY/TDD, thus eliminating the need for specialized equipment.

East Carolina University Project
Attention: Dr. David Lunney or Dr. Robert C. Morrison
Department of Chemistry
East Carolina University
Greenville, NC 27834
(919) 757-6711

Dr. Lunney and Dr. Morrison have developed a system using voice synthesis to translate laboratory data, necessary for experiments, into sound (either speech or music) that can be understood by blind students, enabling them to use scientific instruments independently.

Future Tech: Project Coffee
Attention: John R. Phillipo or Margaret Reed
Oxford High School Annex
Main Street
Oxford, MA 01540
(617) 987-1626; (617) 987-1727

Project Coffee has used existing resources and available revenue to develop an integrated educational program which has been validated by the U.S. Department of Education as a model program. The computer-related curricula intended for all students as well as the disadvantaged/unemployed adults of the community have effected significant changes in the academic progress and employability of those involved.

Helen Keller National Center
Attention: Sr. Bernadette Winn
111 Middle Neck Road
Sands Point, NY 11050
(516) 944-8900

Menus, bus schedules, game rules, medical information, books, and letters are translated into Braille via computer for individual center residents. In addition, deaf-blind staff members have ready access to office communications which are available in both Braille and print. A Braille keyboard allows a blind person to type in on the computer, and a special printer prints out the communication in Braille, thus providing direct interaction between the handicapped person and in-house information systems.

InterLearn, Inc.
Box 342
Cardiff-by-the-Sea, CA 92007

InterLearn has created a variety of educational software, much of it language-oriented, and much of it created with special needs students in mind. The *Computer Chronicles Newswire* combines electronic mail with a child-oriented word-processing system, *The Writer's Assistant*, both to encourage writing and to provide an intriguing context for it— letting the children be reporters for a news network that extends across the nation. The *Network Toolbox* allows students, teachers, and home users to connect their Apple and modem up to information resources like *The Source* or to other electronic mail networks, and to transfer work quickly to other computers, as with the *Newswire*. InterLearn offers consulting services to help groups establish specially tailored networks. Interactive reading and writing software based on InterLearn's *Interactive Text Interpreter* include tools for helping students learn to write expository texts, narrative texts, and poetry. The teacher (and students) can create fully personalized interactive texts for any purpose. A *Spelling Toolbox* uses musical and visual imagery to help students become familiar with the spelling patterns of words and remember the spelling of difficult or irregular words.

Language Development in Deaf Children Using Interactive Videodisc
Dr. Robert K. Lennan, Superintendent
Attention: Rod Brawley/Barbara Peterson
California School for the Deaf
3044 Horace Street
Riverside, CA 92506
(714) 683-8140

The school is providing hearing-impaired children with language and reading experiences in a visual medium. Using visual material on the videodisc, the teacher acts as author, entering text and creating tasks on the Apple II to meet the reading level and special needs of the individual student. The hope is that the hearing-impaired will develop English skills that more closely approximate the skills of their hearing counterparts.

Market Linkage Project for Special Education
LINC Resources, Inc.
1875 Morse Road, Suite 225
Columbus, OH 43229
(614) 263-5462

Begun in 1977 by LINC (see **Training and Resource Centers**), the project provides technical marketing assistance to those developers of products for handicapped learners who have been awarded grants and contracts by the U.S. Department of Education, Office of Special Education.

Michigan State University Artificial Language Laboratory
Computer Science Department
East Lansing, MI 48824
(517) 353-0870

The focus of their research is on developing communication prosthetics, including voice synthesis, for the physically handicapped and training in handwriting and word processing for the blind. The lab publishes *Communication Outlook* (see **Publications: Periodicals**) which provides a forum for individuals interested in the application of technological aids for those who experience communication handicaps as a result of neurological or neuromuscular conditions.

Microcomputers in the Schools—Implementation in Special Education
Attention: Laura Clark
SRA Technologies
901 S. Highland Street
Arlington, VA 22204
(703) 486-0600

In conjunction with the U.S. Department of Education, SRA and the Case Study Institute are involved in a two-year research and information dissemination project to examine and describe the use of microcomputers in special education programs. Some of the issues being looked at are the level and nature of planning involved in the implementation of microcomputers, the emergence of new roles as a function of their introduction, and the relationship between special education and regular education programs as a result of microcomputer use.

Multimedia Access to Microcomputers for Visually Impaired Youth
S.C. Ashcroft, Principal Investigator
Peabody College of Vanderbilt
Department of Special Education
Box 328
Nashville, TN 37203
(615) 322-8165

This project is designed to study visual, auditory, and tactile methods to give visually impaired youth access to microcomputers for curricular prevocational and avocational purposes.

National RETOOL Center
Attention: Josephine Barresi
The Council for Exceptional Children
1920 Association Drive
Reston, VA 22091
(703) 620-3660

With a grant from the U.S. Department of Education, the National RETOOL Center (see **Training and Resource Centers**) is developing, over a three-year period, a series of six software training packages that deal with such topics as selecting software and hardware, adapting software for handicapped children, peripherals, and microcomputer programming for instructional purposes. Also underway is the development of a national network to be called Computer Using Teacher Education Network (CUTE).

Newington Children's Hospital, Department of Speech Pathology
Zane Saunders, Director
181 Cedar Street
Newington, CT 06111
(203) 667-5200

They provide a full augmentative communication assessment and training program for children and adults. In addition, microcomputers are used as a therapy tool with traumatic brain-injured and cerebral palsy patients.

Non-Oral Project, Newcastle School
Attention: Joanne Ligamari
P.O. Box 58
Newcastle, CA 95658-0058
(916) 663-3338

The purpose of the project is to investigate the use of computers to aid communication for non-oral, physically handicapped children ages 3–10 and to assist in the instruction of mathematics and reading. As a management tool for the project, the computer aids in the scheduling and monitoring of handicapped students who are mainstreamed.

Project C.A.I.S.H.
Warren R. Brown, Project Manager
3450 Gocio Road
Sarasota, FL 33580
(813) 355-3567

Project C.A.I.S.H. is a response to the needs of students with handicaps that inhibit their ability to communicate. Microcomputers are used to provide a communication link between students, teachers and instructional materials. The plan is to develop curriculum materials for use by handicapped students and better meet their instructional needs.

Project CAL
Attention: Dr. Dolores Shanahan
Commack Public Schools
Hubbs Administration Building
Box 150
Commack, NY 41725
(516) 493-3574

To promote intra-school learning, high school students instruct K–6th grade students in the basics of computer technology. The Commack School District has also participated with BOCES III to bus gifted/handicapped students to the Commack Computer Lab for CAI in mathematics, language arts, and the fundamentals of BASIC. As a result, Commack has developed COMCAL (COMMACK COMPUTER ACCELERATED LEARNING), a bank of software programs written cooperatively by students (see Chapter 7 of this book for a longer description of Project CAL).

The Rehabilitation Institute of Pittsburgh
Attention: Mata Jaffee, Ph.D.
The Rehabilitation Institute of Pittsburgh
6301 Northumberland Street
Pittsburgh, PA 15217
(412) 521-9000

Intensive summer sessions for young and old people with spina bifida take place here, including training via computerized tutorials which serves to educate them, thus fostering independence for self-care.

The Schneier Communication Unit of the Cerebral Palsy Center
Carol G. Cohen, Director
1603 Court Street
Syracuse, NY 13208
(315) 455-5726

The Schneier Communication Unit is involved in research and development of software and hardware innovations to allow individuals with physical disabilities and severe communication impairments independence and autonomy in daily living at home, school, or work.

Sensory Aids Foundation (SAF)
Attention: Susan H. Phillips
399 Sherman Avenue, Suite 12
Palo Alto, CA 94306
(415) 329-0430

With a grant from the U.S. Department of Education, SAF (see **Associations and Organizations**) is adapting commercially available software for use by the blind. Software is being modified for use on an Apple IIe computer interfaced with a speech synthesizer to speak the information that appears on the video screen.

Shrine School for the Physically Handicapped
Attention: Monte B. Burns/Patricia Huggins
4259 Forest View
Memphis, TN 38118
(901) 795-3930

The Shrine School aids students with physical handicaps in academic subjects and vocational training through the use of unique keyboard error-trapping software, and a variety of alternate interface devices such as switches, paddles, and expanded keyboards.

University of Washington, Seattle
Attention: Dr. Wesley Wilson
Seattle, WA 98195
(206) 543-7039

Dr. Wilson has built a communication system that allows paralyzed nonvocal individuals to use voice synthesis and Morse code.

PUBLICATIONS

A growing number of publications now address issues in education, special needs, and computers. This is due to recent technological advancements and new research findings. This section lists and briefly describes only some of the available books, articles, special reports, bibliographies, magazines, and newsletters. Also included is a bibliography of references from the various chapters of this book.

For a listing of the myriad computer and educational computing publications see the Resources Section of *Practical Guide to Computers in Education*.

Periodicals

Bounty
Joanne Doell, Publisher
17710 Dewitt Avenue
Morgan Hill, CA 95037

This quarterly publication focuses on language/learning disabled students. It includes a special section entitled "Computer Corner," which reviews special education software. Subscriptions are $6 per year.

THE CATALYST
Sue Swezey, Editor
Western Center for Microcomputers in Special Education
1259 El Camino Real—Suite 275
Menlo Park, CA 94025
(415) 326-6997

THE CATALYST offers current information on programs and resources utilizing microcomputers for special education. Subscriptions are available for organizations ($20) and individuals ($12).

Closing the Gap
Budd Hagen, Editor
P.O. Box 68
Henderson, MN 56044
(612) 665-6573

Closing the Gap is a friendly and personal newsletter filled with pertinent information on computers and related programs for people with a variety of disabilities. It offers an information exchange page for readers and schedules workshops (see **Training and Resource Centers**) on the uses of microcomputers in education. Subscriptions are $15 per year for six issues.

Communication Outlook
Artificial Language Laboratory
Computer Science Department
Michigan State University
East Lansing, MI 48824
(517) 353-0870

A quarterly newsletter published jointly by the TRACE Center (see **Training and Resource Centers**) and the Michigan State University Artificial Language Laboratory (see **Promising Practices**), *Communication Outlook* is directed toward individuals interested in technological applications related to communication handicaps.

COPH Bulletin
Committee on Personal Computers and the Handicapped
2030 Irving Park Road
Chicago, IL 60618
(312) 477-1813

This is the official publication of the Committee on Personal Computers and the Handicapped (see **Associations and Organizations**). The purpose of this periodical is to search out, evaluate, and share personal computer information important to those with physical disabilities.

The First International Review of Special Education Technology
Judy Smith-Davis
Counterpoint Communications Co.
750 McDonald Drive
Reno, NV 89503
(702) 747-7751

The first edition of what will be an annual publication covers a wide range of topics. Sponsored by the Association for Special Education Technology (see **Associations and Organizations**), it deals with microcomputers and related electronic technology. Topics included are: CAI, CBI, Telecommunications Applications, Electronic Instructional Aids for Persons with Sensory and Other Impairments, and Technology for the Future.

Focus on Exceptional Children
Love Publishing Company
1777 South Bellaire Street
Denver, CO 80222
(303) 757-2579

Published nine times during the school year, the articles are geared toward teachers, special educators, curriculum specialists, administrators, and others concerned with the education of exceptional children. Two issues, October 1981 and February 1983, focus on microcomputers in this field. The subscription rate is $18 per year.

GA-SK Newsletter
Barry Strassler
National Association of the Deaf
814 Thayer Avenue
Silver Spring, MD 20910
(301) 589-3006

This newsletter is published by Telecommunications for the Deaf, Inc. (see **Associations and Organizations**) and is available to members for $10 per year. Also available through TDI is an *International Telephone Directory for the Deaf*, which lists TDD/TTY numbers.

Journal of Educational Technology Systems
Society for Applied Learning Technology
Baywood Publishing Co., Inc.
120 Marine Street
Box D
Farmingdale, NY 11735

Published quarterly, this journal may be obtained at a subscription rate of $25 per year.

Journal of Learning Disabilities
Gerald M. Sent, Ph.D.
1331 E. Thunderhead Drive
Tucson, AZ 85718
(602) 297-2842

The *Journal of Learning Disabilities* recognizes the significance of microcomputer-based education for learning disabled students. It includes a section called "Computers in the Schools," which is devoted to microcomputer applications, and it offers reviews of courseware. It is involved with a project to encourage major producers of hardware to develop compatible software appropriate for the LD population.

Journal of Special Education Technology
Managing Editor
Exceptional Child Center
UMC-68
Utah State University
Logan, UT 84322

The *Journal of Special Education Technology* is a quarterly publication of the Association for Special Education Technology (ASET) (see **Associations and Organizations**). It provides information on recent research and innovative practices regarding the application of educational technology in the education of handicapped children.

Sensory Aids Technology Update
Sharon Connor, Editor
399 Sherman Avenue, Suite 12
Palo Alto, CA 94306
(415) 329-0430

A publication of the Sensory Aids Foundation (see **Associations and Organizations**), this monthly newsletter focuses on practical applications and advances made in technology for both hearing and visual impairments. The subscription rate is $30 per year.

Update
Sally Bulfod, Carol Daniels
LINC Resources, Inc.
1875 Morse Road, Suite 225
Columbus, OH 43229

Update provides information about commercial products for special education and is a good source of information about computer technology and its applications for handicapped individuals.

Special Issues of Magazines

BYTE Magazine, August 1982, Vol. 7, No. 8
BYTE Publications
70 Main Street
Peterborough, NH 03458

This issue is devoted to the use of Logo.

BYTE Magazine, September 1982, Vol. 7, No. 9
BYTE Publications
70 Main Street
Peterborough, NH 03458

This issue focuses on computers and the disabled.

Classroom Computer Learning, October 1983
Pittman Learning Co.
19 Davis Drive
Belmont, CA 94002
(415) 592-7810

This issue focuses on computers and special needs.

Computer, January 1981, Vol. 14, No. 1
IEEE Computer Society
10662 Los Vaqueros Circle
Los Alamitos, CA 90720
(714) 821-8380

This issue is devoted to computing and the handicapped.

Computer Decisions, October 1976, Vol. 8, No. 10
Hayden Publishing Company, Inc.
50 Essex Street
Rochelle Park, NY 07662
(201) 843-0550

This issue is devoted to computers and the handicapped.

The Computing Teacher, February 1983, Vol. 10, No. 6
135 Education
University of Oregon
Eugene, OR 97403
(503) 686-4414

A publication of The International Council for Computers in Education (see **Associations and Organizations**), this issue is devoted to computers and special needs. This publication is issued monthly nine times during the academic year, with a subscription rate of $16.50 per year.

IEEE Micro, June 1983
Jim Aylor, Editor
IEEE Computer Society
P.O. Box 80452
Worldway Postal Center
Los Angeles, CA 90080
(703) 367-3477

This issue is devoted to computers and the handicapped.

Reports and Articles

Access to Computers for the Physically Handicapped
Carol B. Fusca, Marketing
Prentke Romich Co.
8769 Township Road, 513
Shreve, OH 44676
(216) 567-2906

The Prentke Romich Company has worked to develop products that enable the physically handicapped to have access to standard hardware and software. This publication describes their products and specific applications as well as listing resources and information about other companies' products.

Computer Systems for Special Educators
Kirk Wilson
Learning Tools
686 Massachusetts Avenue
Cambridge, MA 02139
(617) 864-8086

This report is intended for special education administrators who plan to use computers to aid in program management. It describes the stages of selecting and implementing a system and outlines specific applications of computers in special education.

Demonstration of the Use of Computer Assisted Instruction with Handicapped Children: Final Report
Bolt Beranek and Newman, Inc.
10 Moulton Street
Cambridge, MA 02138
(617) 491-1850

This report contains the results of a study, sponsored by the U.S. Department of Education, designed to demonstrate the usefulness of a computer aided educational environment for deaf children.

The Evaluation and Cultivation of Spatial and Linguistic Abilities in Individuals with Cerebral Palsy, LOGO Memo #55, 1980
LOGO Group
MIT
545 Technology Square
Cambridge, MA 02139

This is a summary of present and future research being conducted by the Logo Group in using computers to teach children with cerebral palsy. The information includes the use of computer-based systems to enhance development of language and spatial skills, the development of diagnostic tools, and the application of these areas to theories of cognitive development.

Final Report/ILIAD: Interactive Language Instruction Assist for Deaf
Bolt Beranek and Newman, Inc.
10 Moulton Street
Cambridge, MA 02138
(617) 491-1850

ILIAD is a computer system designed to aid deaf students in language development by providing tutorials on the production and comprehension of written English sentences. The focus of ILIAD is not on what the student learns but how s/he learns and explores a wide range of simple to complex language forms.

Finding a Voice, Talking Turtle: NOVA Transcripts
NOVA Transcripts
P.O. Box 322
Boston, MA 02134
(617) 492-2777

Finding a Voice, first broadcast on February 7, 1982, was written and narrated by cerebral palsy victim Dick Boydell who cannot speak and communicates by means of a special typewriter that he operates with his foot. The transcript can be ordered at bulk (20 at $1.50 each) or individual ($3) rates. *Talking Turtle*, first broadcast on October 25, 1983, shows Logo being used in a wide variety of settings, including work with physically handicapped children.

Implications of Video Disc Microcomputer Instructional Systems for Special Education
Ron Thorkildsen, Joseph G. Williams, Warren Bickel
Utah State University
Logan, UT 84322
(801) 750-1999

This report describes a research project conducted at Utah State University's Exceptional Child Center that used videodisc technology to develop the MCVD (Microcomputer Video-Disc system). The system is intended to provide individualized instruction for moderately retarded learners through the storage and presentation of audio-visual information. PILOT (an authoring language) was modified to interact with the videodisc and the touch panel to allow the learner to respond to audio-instruction and an associated visual image by touching the television screen.

Instructional Technology for Special Needs
Information Services
Ministry of Education
Parliament Buildings,
Victoria, B.C. V8V 2M4

This discussion paper describes current research dealing with technology and the handicapped. The areas covered include: sight, hearing, intelligence, motor coordination, and communication. Speculation on future applications of technology to people with special needs is offered.

Learning Disabled Students and Computers: A Teacher's Guidebook
Merrianne Metzger, David Ouellette, and Joan Thormann
International Council for Computers in Education, 1983
135 Education
University of Oregon
Eugene, OR 97403
(503) 686-4414

This booklet is a nontechnical introduction to simple and practical educational applications of microcomputers with learning disabled students.

Logo: A Computer Environment for Learning Disabled Students; *The Computing Teacher*, Vol. 8, No. 5, 1980–1981
135 Education
University of Oregon
Eugene, OR 97403
(503) 686-4414

This article describes the work at the Logo Group with learning disabled students in grades 5–8 and gives a preview of proposed projects.

Observations on Report, "Technology and Handicapped People"
Alan M. Hofmeister, Dean
Utah State University
School of Graduate Studies
Logan, UT 84322

Dr. Hofmeister is a leader in the field of technology for handicapped people and has been largely concerned with legislation and regulations of technology in relation to handicapped individuals. His report, entered as testimony before the Senate Committee on Labor and Human Resources and the House Subcommittee on Science, Research, and Technology, September 29, 1982, would be beneficial to those developing programs or seeking funds for programs and/or equipment.

Proceedings of the Johns Hopkins First National Search for Applications of Personal Computing to Aid the Handicapped
IEEE Computer Society
P.O. Box 80452
Worldway Postal Center
Los Angeles, CA 90080
(714) 821-8380

In November 1980, the Johns Hopkins University Applied Physics Laboratory launched a nationwide search for privately developed applications of personal computing to aid the handicapped. Reported in this publication is a sampling of the top regional entries. The abstracts are grouped by the following handicap categories: hearing, speech and language; learning disabilities and mental retardation; movement, neuromuscular, and neurological; vision; and nonspecified. This 304-page book is available for $22.00 ($16.50 for IEEE members).

Research in Applying Personal Computers to Telecommunications and Education of the Deaf
Howard Cunningham
AMRAD
P.O. Drawer #6148
McLean, VA 22106

This is a project report prepared for the Department of Education describing AMRAD's (Amateur Radio Research and Development Corporation) research in applied uses of personal computers and telecommunications for the deaf (see **Promising Practices**).

A Resource Guide: Personal Computers for the Physically Disabled
Apple Computer Inc.
20525 Mariana Avenue
Cupertino, CA 95014
(408) 996-1010

The Apple Computer Resource Guide is available upon request and offers detailed descriptions of what personal computers can do for handicapped people. A bibliography and list of personal contacts are included.

Technology and Handicapped People
John H. Gibbons, Director
Office of Technology Assessment
Congress of the United States
Washington, DC 20510

The Office of Technology Assessment conducted a study of the development and use of technology for handicapped individuals to examine and evaluate the performance of services and programs. According to John Gibbons, "the appropriate application of technologies to diminishing the limitations and extending the capabilities of disabled and handicapped persons is one of the prime social and economic goals of public policy."

Telecommunications for the Deaf: A History of Technological Development of Devices to Aid the Deaf
Rudolph Auslander
Mill Neck School for the Deaf
P.O. Box 12
Mill Neck, NY 11765
(516) 922-4100

The purpose of Mr. Auslander's monograph is to explain the technological developments of the past fourteen years which have provided new hope for the deaf and hearing impaired. He describes the potential of microcomputers to provide the deaf with equipment capable of communicating with both existing deaf teletypewriters and computer networks, and he discusses hardware requirements as well as costs of setting up a system for home use. As a parent of a deaf child, Mr. Auslander has done extensive research and has several smaller reports on microcomputers for deaf communications.

Using Computers to Assist Learning Disabled High School Students
Linda K. Reiss
High School of Commerce
300 Rochester Street
Ottawa, Ontario, CA K1R 7N4

This report prepared for the Ottawa Board of Education describes computer aided instruction for learning disabled secondary students in the United States and Canada. It also addresses the issues of software effectiveness, Logo, special education management, and the training of learning disabilities specialists in the use of computers with students.

Bibliographies

Bibliography of Logo Memos
Educational Computing Group
Laboratory of Computer Science
545 Technology Square, Room 908
Cambridge, MA 02139

Logo is a computer programming language that has been used extensively with children who have special needs. Reports and papers describing this work are available and may be ordered from the above address.

Computer Technology for the Handicapped in Special Education and Rehabilitation: A Resource Guide
International Council for Computers in Education
135 Education
University of Oregon
Eugene, OR 97403
(503) 686-4414

This guide includes annotated citations from over thirty-nine journals. The entries focus on computer assisted instruction and computer managed instruction. Author and subject indexes are also provided.

Microcomputers in Special Education
Stephen Krasner
Special Education Resource Center
275 Windsor Street
Hartford, CT 06120
(203) 246-8514

This bibliography, prepared in April 1982, lists journal articles relevant to computers and the handicapped.

Technology in Special Education Instruction: Annotated Bibliography #18
JWK International Corporation
Project EduTech
7617 Little River Turnpike
Annandale, VA 22003

Included here are annotated references related to the application of technology to special education.

Bibliography of References in This Book

Chapter 1

Goldenberg, E. P., *Special Technology for Special Children.* Baltimore: University Park Press, 1979.
Goldenberg, E. P., "Flexible High Bandwidth Communication for Motorically Impaired Persons," *Proceedings of the Johns Hopkins First National Search for Applications of*

Personal Computing to Aid the Handicapped, October 31, 1981. Worldway Postal Center, Los Angeles: IEEE Computer Society Press.

Goldenberg, E. P., "Computers in the Special Education Classroom: What Do We Need and Why Don't We Have Any?" In Mulick, J. A., and Mallory, B. L., eds., *Transitions in Mental Retardation: Advocacy, Technology and Science*. Ablex Press, Norwood, NJ: Ablex Publishing Co., 1984.

Howe, J. A. M., "Some Roles for the Computer in Special Education," DAI Research Paper No. 126, Department of Artificial Intelligence, University of Edinburgh, Edinburgh, Scotland.

La, W. H. T., Koogle, T. A., Jaffe, D. L., and Leifer, L. G., "Toward Total Mobility: An Omnidirectional Wheelchair," *Proceedings of the Fourth Annual Conference on Rehabilitation Engineering*, Washington, D.C., pp. 75–77, 1981.

LeBlanc, M., "Systems and Devices for Nonvocal Communication," In Bleck, E. E., and Nagel, D. A., eds., *Physically Handicapped Children—A Medical Atlas for Teachers*, Second Edition. New York: Grune and Stratton, 1982, pp. 159–169.

Loebl, D., "Microcomputer as a Tool for Testing and Teaching Developmentally Impaired Children Color and Shape Discrimination." Unpublished doctoral dissertation, Boston University, Boston, MA, 1982.

Papert, S., *Mindstorms: Computers, Children, and Powerful Ideas*. New York: Basic Books, 1981.

Chapter 2

Goldenberg, E. P., 1984. (See Chapter 1 above.)

Riel, M., "Investigating the System of Development: The Skills and Abilities of Dysphasic Children," University of California, San Diego, Technical Report # CHIP-115, February, 1983.

Riel, M., "Education and Ecstasy: Computer Chronicles of Children Writing Together," *Quarterly Newsletter of the Laboratory of Comparative Human Cognition* 5:3, 1983.

Chapter 3

Cory, L. W., Viall, P. H., and Walder, R., "Computer Assisted Communication," In Mulick, J. A., and Mallory, B. L., eds., *Transitions in Mental Retardation: Advocacy, Technology and Science*. Norwood, N.J.: Ablex Publishing Co., 1984.

Gengel, R., "Research with Upton's Visual Speechreading Aid," In Pickett, J. M., ed., *Proceedings of the Research Conference on Speech Processing Aids for the Deaf*. Washington, D.C.: Gallaudet College, 1977.

Goldenberg, E. P., 1979. (See Chapter 1 above.)

LeBlanc, M., 1982. (See Chapter 1 above.)

Levin, J., "Microcomputers and Interactive Communication Media: An Interactive Text

Interpreter," *Quarterly Newsletter of the Laboratory of Comparative Human Cognition*, 4(2):34–37, 1982.

Levin, J., Boruta, M., and Vasconcellos, M., "Microcomputer Based Environments for Writing: A Writer's Assistant," In Wilkinson, A. C., ed., *Classroom Computers and Cognitive Science*. New York: Academic Press, 1983.

Levin, J., Riel, M., Rowe, R., and Boruta, M., "Muktuk Meets Jacuzzi: Computer Networks and Elementary School Writers," In S. Freedman, ed., *The Acquisition of Written Language: Revision and Response*. Norwood, NJ: Ablex Publishing Co., 1984.

McNaughton, S., "Augmentative Communication System: Blissymbolics," In Bleck, E. E., and Nagel, D. A., eds., *Physically Handicapped Children—A Medical Atlas for Teachers*, Second Edition, New York: Grune and Stratton, 1982, pp. 146–154.

Pickett, J. M., Gengel, R., and Quinn, R. "Research with the Upton Eyeglass Speechreader," In Levitt, H., Pickett, J. M., and Houde, R., eds., *Sensory Aids for the Hearing Impaired*. New York: IEEE Press, distributed by John Wiley and Sons.

Prizant, B., "An Analysis of the Functions of Immediate Echolalia in Autistic Children," Unpublished doctoral dissertation, State University of New York, Buffalo, 1978.

Sharples, M., "A Computer Written Language Lab," *Computer Education*, No. 37, 1981, pp. 10–12. (Also available as DAI Research Paper No. 134, from Department of Artificial Intelligence, University of Edinburgh, Edinburgh, Scotland.)

Chapter 4

Ballas, M. S., "Computer Drill and Practice Make the Grade," *ETS Developments* 28:1, May 1982.

Carpenter, T. P., Corbitt, M. K., et al., "Results and Implications of the Second NAEP Mathematics Assessment: Elementary School," *Arithmetic Teacher* 27:8, April 1980.

Carter, R., "The Complete Guide to Logo," *Classroom Computer News* 3:5, April 1983.

Dugdale, S., *Using the Computer to Foster Creative Interaction Among Students*, CERL Report e-9. Computer Based Education Research Laboratory, University of Illinois, Urbana, October 1979.

Friel, S., "Lemonade's the Name, Simulation's the Game," *Classroom Computer News* 3:3, January/February 1983.

Hawkins, D., "Messing About in Science," *The Informed Vision: Essays on Learning and Human Nature*. New York: Agathon, 1974.

Kelman, P., et al., *Computers in Teaching Mathematics*. Reading, MA: Addison-Wesley, 1983.

Kruteskii, V. A., *The Psychology of Mathematical Abilities in School Children*. Chicago: The University of Chicago Press, 1976.

National Council of Supervisors of Mathematics, "Position Paper on Basic Skills," *Arithmetic Teacher* 25:2, October 1977.

Papert, S., 1981. (See Chapter 1 above.)

Papert, S., et al. "Final Report of the Brookline Project, part II: Project Summary and Data Analysis," LOGO Memo 53, Educational Computing Group, MIT.

Post, T. R., "Fractions: Results and Implications from National Assessment," *Arithmetic Teacher* 28:9, May 1981.
Schwartz, J., "The Semantic Calculator: Solving the Word-Problem Problem," *Classroom Computer News* 2:4, March/April 1982.
Sells, L. W. "The Mathematics Filter and the Education of Women and Minorities," In Fox, Brody, and Tobin (eds.), *Women and the Mathematical Mystique*. Baltimore: Johns Hopkins University Press, 1980.
Suppes, P., "The Uses of Computers in Education," *Scientific American* 215:6, July 1966.
Watt, D., "Final Report of the Brookline LOGO Project, part III: Profiles of Individual Student's Work," LOGO Memo 54, Educational Computing Group, MIT.
Watt, D., *Learning with Logo*. New York: McGraw-Hill, 1983.
Weir, S., Russell, S. J., and Valente, J. "LOGO: An Approach to Educating Disabled Children," *Byte*, September 1982.

Chapter 5

Goldenberg, E. P., 1979. (See Chapter 1 above.)
Meichenbaum, D., "Self-Instructional Methods," In Kanfer, F. and Goldstein, A., eds., *Helping People Change*. New York: Pergamon Press, 1975.
van Lint, J., *My New Life*. San Diego: Neyenesch Printers, 1975.
Wier, S. and Emanuel, R., "Using Logo to Catalyze Communication in an Autistic Child," D.A.I. Research Report No. 15, Department of Artificial Intelligence, University of Edinburgh, Edinburgh, Scotland, 1976.

Chapter 6

Ashlock, R. B., *Error Patterns in Computation*. Columbus, OH: Charles E. Merrill, 1976.
Barclay, T., "Buggy: Outfitting for the Great Error Hunt," *Classroom Computer News* 2:4, March/April 1982.
Emmerichs, J., *Superwumpus*, Peterborough, NH: *Byte Publications*, 1978.
Engel, B. S., *Informal Evaluation*. North Dakota Study Group in Evaluation, Center for Teaching and Learning. Grand Forks: University of North Dakota, March 1977.
Eric Clearinghouse on Handicapped and Gifted Children, *Computers for Special Education Management: Progress, Potential, and Pitfalls*. Reston, VA: Council for Exceptional Children.
Hoffman, B., *The Tyranny of Testing*. New York: Crowell-Collier, 1962.
Houts, P. L., ed., *The Myth of Measurability*. New York: Hart, 1977.
Hutchens, M., "Learning Disabilities: A Futuristic Approach," *Journal of Learning Disabilities* 13:9, November 1980.

Yob, G., "Wumpus 2." In D. Ahl (ed.), *The Best of Creative Computing.* Morristown, NJ: Creative Computing Press, 1979.

Chapter 7

Goldenberg, E. P., 1984. (See Chapter 1 above.)
Harvey, B., "Why Logo? " *BYTE,* 7(8):163, August, 1982.
Harvey, B., "Stop Saying 'Computer Literacy'," *Classroom Computer News,* 3(6), May/June, 1983.
Nashel, D. J., Korman, L. Y., and Bowman, J. O., "Radiation Hazard of Video Screens," *New England Journal of Medicine,* 307(14):891, 1982.
Russell, S. J., "Had We But World Enough and Time: Logo and Special Education," *Classroom Computer Learning,* 3(9), October, 1983.
Tucker, M. S., "Computers in Our Schools: A Policy Perspective." Paper available from Project on Information Technology and Education, Suite #301, 1011 Connecticut Avenue, NW,Washington, DC 20036, 1983.
Tucker, M. S., "Planning for Computers in Schools: A Stitch in Time Saves Nine," *Theory Into Practice* (a journal of The Ohio State University), November, 1983.
Zazula, J., and Foulds, R., "A Developmentally Appropriate Mobility Device for a Child with Multiple Congenital Limb Deficiencies, *Archives of Physical Medicine and Rehabilitation,* Vol. 64, March, 1983, pp. 137–139.

Books on Computers, Education, and Special Needs

Budoff, M., *Microcomputers in Special Education.* Cambridge, MA: Ware Press, 1984.

Cleary, A., Mayes, T., and Packham, D., *Educational Technology: Implications for Early and Special Education.* London: John Wiley and Sons, 1976.

> Published early in the history of computer use in education and specifically with the handicapped, it presents an introduction to the theory underlying the use of technology in special education. Included is information about the first teaching machines, video technology, and the early use of computers.

Coburn, P., Kelman, P., Roberts, N., Snyder, T., Watt, D., and Weiner, C., *Practical Guide to Computers in Education.* Reading, MA: Addison-Wesley, 1982.

> This is the first book in the *Computers in Education* series. It is designed to assist educators in exploring the potential of computers in education—a basic primer.

Through vignettes and pragmatic suggestion, the authors relate nontechnical information about such issues as introducing the computer to the school, limits and potentials of microcomputers, using the computer creatively, and finding effective resources.

Geoffrion, L., and Geoffrion, O., *Computers and Reading Instruction*. Reading, MA: Addison-Wesley, 1983.

This comprehensive book, part of the *Computers in Education* series by Addison-Wesley, presents ways computers can be used effectively for teaching reading. Each chapter is filled with examples of useful computer programs, as well as ideas for future development by the classroom teacher and the industry.

Goldenberg, E. P., *Special Technology for Special Children: Computers to Serve Communication and Autonomy in the Education of Handicapped Children*. Baltimore, MD: University Park Press, 1979.

With numerous case studies and a psychological orientation, this book examines several communication impairments, including autism, deafness, and cerebral palsy, as they are elucidated by and influenced by the use of computers as informational prostheses. The volume cites extensive research and provides a psychological foundation for the use of computers in educating exceptional children.

Goldenberg, E. P., Carter, C., and Russell, S., with Stokes, S., Sylvester, J., and Kelman, P., *Computers, Education and Special Needs*. Reading, MA: Addison-Wesley, 1984.

A practical guide to using computers with special needs students, it contains a detailed resource list including sources of hardware and software, and promising programs and practices currently underway.

Kelman, P., Bardige, A., Choate, J., Hanify, G., Richards, J., Roberts, N., Walters, J., Tornrose, M., *Computers in Teaching Mathematics*. Reading, MA: Addison-Wesley, 1983.

This book presents a wealth of ideas, including the basic information mathematics teachers need to introduce computers into their schools. It explores the potential for a curriculum revolution in mathematics, in addition to emphasizing traditional uses of the computer in teaching mathematics. It addresses what to expect, and what not to expect, from a computer, as well as how to use it as a teaching tool. There are specific chapters on computer-assisted instruction, programming, graphics, problem-solving, computer science, and designing computer-based units and courses.

Taber, F., *Microcomputers in Special Education: Selection and Decision Making Process*. Reston, VA: The Council for Exceptional Children, 1983.

This book provides an overall introduction to microcomputers in education, addressing in particular the uses in special education. Each chapter includes a bibliography and sources of further information.

Books on Particular Disabilities

The following books are ones which we have found to provide valuable insight into the particular disability discussed.

Bleck, E. E., and Nagel, D. A., eds., *Physically Handicapped Children — A Medical Atlas for Teachers*, Second Edition. New York: Grune and Stratton, 1982, pp. 159–169.

> Covers both common and low-frequency handicaps (varying from asthma, heart defects, and allergies through motoric and sensory handicaps) important for the classroom teacher to know about. While not primarily devoted to issues of pedagogy, this excellent resource provides information to help the classroom teacher understand how the child's condition may affect classroom participation and learning. Includes some educational interventions.

Edgerton, R. B., *Mental Retardation*. Cambridge, MA: Harvard University Press, 1979.

> Current, very readable introduction to mental retardation, its varieties, some of its causes, and outlooks.

Farnham-Diggory, Sylvia, *Learning Disabilities*. Cambridge, MA: Harvard University Press, 1978.

> Very clear, highly readable presentation of enlightened views on several of the learning disabilities.

Featherstone, H., *A Difference in the Family — Life with a Disabled Child*. New York: Basic Books, 1980.

> Insightful, informed, and literate firsthand account of the effects on the family of a child with multiple severe handicaps.

Gibson, D., *Down's Syndrome — The Psychology of Mongolism*. New York: Cambridge University Press, 1978.

> A thorough, informative, non-prescriptive, developmentally sound treatment of Down's syndrome, debunking old myths, and presenting new prospects. It provides medical, social, psychological, historical/political views, and educational interventions.

Mindel, E. D., and Vernon, M., *They Grow in Silence — The Deaf Child and His Family*. Silver Spring, MD: National Association of the Deaf, 1971.

> A psychiatrist and psychologist present the psychological and educational implications of growing up deaf.

Park, C. C., *The Siege — The First Eight Years of an Autistic Child*. Boston: Atlantic-Little, Brown, 1967.

A beautifully written firsthand account of family life with an autistic girl, this book presents a family perspective to complement research views.

van Lint, J., *My New Life*. San Diego: Neyenesch Printers, 1975.

A firsthand account of partial recovery from an accident that left the author paralyzed and without speech. This provides an important picture of the psychology of the loss of communication.

Wing, L., ed., *Early Childhood Autism*, Second Edition. London: Pergamon Press, 1976.

This well-edited collection of articles presents a psychology and pedagogy for work with autistic children.

ON LINE NETWORKS AND DATABASES

Through the combination of microcomputer technology and the telephone system, individuals or organizations can subscribe to databases and information retrieval systems to get ball scores, stock quotes, personal mail—they can even read a novel or chat with another user or group of users.

Networks have lessened the isolation of individuals with severe handicaps and have given them access to a larger world of information than ever before. New opportunities for further academic study or careers in business are now possible because of greatly increased access to resources and people. Networks also permit those concerned with education and special needs to keep in touch with each other and to keep up with the latest developments in the field.

Listed are network systems in operation at the time of publication. Others are being developed, and new branches established every day. The cost remains prohibitive for many, but attempts to lower rates and ensure access are beginning to have an effect on company policies and legislation.

Access Line
J. Roger Cicchese
P.O. Box 280
Boston, MA 02113
(617) 367-8712

Access Line, sponsored by Access to Life, Inc., is a telephone information tape with information on current developments in science, technology, medicine, and media to consumers with special needs. It runs 24 hours a day, 7 days a week, and is updated on Mondays. The phone number of the information tape is (617) 666-9210.

Assistive Device Database System (ADDS)
American International Data Search, Inc.
2326 Fair Oaks Boulevard, Suite C
Sacramento, CA 95825
(916) 925-4554

ADDS provides up-to-date information on adaptive equipment, training programs, service agencies, and resource persons available to individuals with handicaps. ADDS contains information not only on commercially available devices, but on custom-made devices as well.

DEAFNET
East Coast:
Mary Robinson
Deaf Communications Institute
Deaf Community Center
Framingham, MA 91701
(617) 875-3617; (617) 875-0354

West Coast:
Teresa Middleton
DEAFNET SRI International
333 Ravenswood Avenue
Menlo Park, CA 94025
DEAFNET Inquiry lines
TDD (415) 326-1802
Voice (415) 859-2236

DEAFNET provides the means of sending and receiving messages electronically by computer for the deaf. An electronic bulletin board provides information about educational and career opportunities for the deaf as well as a want ad service, medical tips, movie listings, and social events.

Electronic Information Exchange System (EIES)
Murray Turoff, Director
Computerized Conferencing and
Communication Center
323 High Street
Newark, NJ 07102
(201) 641-5321

Under a grant from NSF, the New Jersey Institute of Technology is investigating use of EIES to allow the homebound and severely disabled to communicate with one another and others. Already, homebound senior citizens and children with cerebral palsy are using the system to communicate with others who have access to the exchange.

ERIC (Educational Research Information Center)
The Council for Exceptional Children
1920 Association Drive
Reston, VA 22091
(703) 620-3660

A federally funded system, ERIC collects literature on all aspects of education, including special education. It is accessible via computer network.

Handicapped Education Exchange (HEX)
Richard Barth
11523 Charleton Drive
Silver Spring, MD 20902
(301) 681-7372

HEX is a computerized bulletin board available through the public telephone network. Hearing or deaf people can use HEX to communicate with each other and to exchange ideas about technological aids for the handicapped.

Microcomputer Education Applications Network (MEAN)
Charles L. Blaschke, President
Education TURNKEY Systems, Inc.
256 North Washington Street
Falls Church, VA 22046-4544
(703) 536-2310

A division of TURNKEY (see **Associations and Organizations**), it was created to facilitate the development and dissemination of information concerning microcomputer applications in special education instructional management and administration.

SECTOR: Special Education Computer Technology On Line Resources
Kim Allard or Bob Reid
Utah State University
Logan, UT 84322

Utah State University (see **Training and Resource Centers**) has developed a database of 400 references related to computer use in special education and a database for reviews of microcomputer software for special education teachers. Anyone can request a search for information about computer uses or special hardware and software by writing SECTOR.

SpecialNet
National Association of State Directors of Special Education
1201 16th Street, NW, Suite 610, E
Washington, DC 20036
(202) 822-7933

SpecialNet is available to anyone who has access to a computer terminal or microcomputer. Organizations and individuals can subscribe to SpecialNet for $200 per year and have access to information about special education topics via electronic bulletin boards that are updated daily.

SOFTWARE

The development of any software is a long and costly procedure, and the market for specialized software is too small for most companies to make that investment. Consequently, there is not yet available much software created for special needs of certified value. Recently, however, companies and individuals have developed, through plan or necessity, creative software that invites the user to interact with the computer and gain a sense of direction and control. Others are modifying existing software, including video games, especially to meet the needs of handicapped students. Therapists have found that games provide a recreational environment that motivates patients to exercise and socialize.

Listed are primarily those programs designed specifically for people with special needs, as well as a few general programs referred to in this book. For descriptions of a variety of quality software appropriate for any student regardless of handicaps, see the Resources Sections of other titles in Addison-Wesley's series on *Computers in Education*, especially *Computers in Teaching Mathematics* and *Computers and Reading Instruction*.

We have made no attempt at an exhaustive listing; any such attempt would be a book in itself and obsolete by the time it was published. Instead we try merely to indicate the range of available software with a representative sampling.

We strongly recommend that the user preview software or see it in use before deciding which program is best for a particular need. It is also desirable, if possible, to talk with other users. Their experience and practice could be the best resource at this time of rapid change.

Selected Software for Special Needs

Braille-Edit
Raised Dot Computing
310 S. 7th Street
Lewisburg, PA 17837
(717) 523-6739

Braille-Edit is a versatile word processing system which combines editing in print, voice, and Braille. Both blind and sighted can use the program to prepare their documents. Two

Braille translators are included. The program supports a wide variety of voice and Braille devices. The cost is $300.

CALL (Computer Aided Language Learning)
Bilingual Publications and Computer Services (BIPACS)
33 W. Walnut Street
Long Beach, NY 11561
(212) 685-3459

This software package provides drill and practice and emphasizes pronunciation, reading, spelling, grammar, vocabulary, and fluency. Graphics and a digitized voice that speaks the words are included as is a management system that keeps a record of student responses. The cost is $99.00 with voice; $49.50 without voice.

Edmark Reading Program for the Apple II
Edmark Associates
P.O. Box 3903
Bellevue, WA 98009
(206) 746-3900

The Edmark Reading Program is a beginning reading and language development program for nonreaders, developed with special education classrooms in mind.

Handicapped Typewriter
Rocky Mountain Software, Inc.
1038 Hamilton Street
Vancouver, BC V6B 2R9
(604) 681-3371

This disk is particularly useful for severely physically handicapped individuals who are unable to use a keyboard. Designed for the Apple computer and a Silentype printer, the keyboard is controlled by a single switch. A picture of the keyboard appears on the screen and character selection is made via the scanning cursor. Version 2.0 includes a word and phrase dictionary and a calculator, and costs $99.00. Version 3.0 includes a telephone directory, a dialing and answering service that uses a loudspeaker phone, and an environmental controller, and costs $199.00.

The Linguist
Leslie Hornung, Director of Public Relations
Synergistic Software
830 North Riverside Drive, Suite 201
Renton, WA 98055
(206) 226-3216

The Linguist is a foreign language translation and tutorial software program for the Apple II. It is designed to be used by those wishing to learn English, and allows entry of words, translations, definitions, phrases, and phonetic pronunciations. This program also contains a testing, scoring, and storage system. The cost is $40.00.

NUMBERS (Nemeth Users' Mathematical Braille Effortless Reproduction System)
Raised Dot Computing
310 S. 7th Street
Lewisburg, PA 17837
(717) 523-6739

Both text and mathematical equations entered from a Braille device can be printed out in English on a special dot matrix printer with the use of this program. The cost is $200.00.

Peachy Writer
Dr. Carl Rutledge
Cross Educational Software
P.O. Box 1536
Ruston, LA 71270
(318) 255-8921

This is an easy-to-use text editor that provides large, double-size text on the screen and, with an Epson printer, may print also in double size. Developed for use on an Apple II, the cost is $24.95.

Software Automatic Mouth (SAM)
Don't Ask Computer Software
2265 Westwood Boulevard, Suite B-150
Los Angeles, CA 90064
(213) 379-8811

SAM is an all-software speech synthesizer for the Apple and Atari microcomputers. The disk is capable of producing speech, and has an unlimited vocabulary and fully adjustable pitch and rate. Although *SAM* understands a phonetic spelling system, the disk comes with *Reciter*, an English text-to-speech conversion program, so that users can type in ordinary English spellings of words. *SAM* also allows the user to add speech to BASIC programs. The cost for the Atari version is $59.95, for the Apple II/II+ $124.95.

Special Needs, Volume I (Spelling)
Minnesota Educational Computer Consortium
2520 Broadway Drive
St. Paul, MN 55113
(612) 376-1118

This program drills handicapped students on frequently misspelled primary and intermediate words. The students input answers via the game buttons, the game turn knobs, or any key on the keyboard. The instructor is able to change the words and sentences. The cost is $30.00.

Talking Screen Textwriter
School and Home Courseware, Inc.
1341 Bulldog Lane, Suite C
Fresno, CA 93710
(209) 227-4341

This program combines a word processing program with a speech synthesizer. The learner receives visual and audio presentations of letters, words, and paragraphs.

Visible Speech for the Hearing Impaired
Software Research Corporation
University of Victoria
P.O. Box 1700
Discovery Park
Victoria, BC V8W 2Y2
(604) 477-7246

This software, designed for speech training, graphically displays pitch, rhythm, and amplitude on the screen in response to speech signals picked up by a microphone. Responses are stored and may be recalled and compared to target patterns. The cost is $675.00.

Voice-Based Learning System (VBLS)
Sterling Swift Publishing Co.
1600 Fortview Road
Austin, TX 78704
(512) 444-7570

VBLS is an authoring system that allows students to speak to their computer in any spoken language. The system requires the Scott Instruments voice-entry terminal. The cost is $99.95.

Software Mentioned in This Book

Language Arts and Communications

The Bank Street Writer
Scholastic, Inc.
902 Sylvan Avenue
Englewood Cliffs, NJ 07632
(212) 505-3000

This is an easy-to-use, inexpensive word processor that is specifically designed for youngsters in grades 4 through 12. The instructions are always in full view on the screen, and directions are given each step of the way. The cost is $95.00.

CARIS (Computer Animated Reading Instruction System)
Britannica Computer-Based Learning
425 North Michigan Avenue
Chicago, IL 60611

This program uses animated cartoons to introduce the reading and spelling of an initial vocabulary. Teachers have considerable flexibility in selecting and labeling pictures and actions.

Story Machine
Spinnaker Software
215 First Street
Cambridge, MA 02142

Children create short stories involving up to four separate objects and eleven different actions. The program generates a cartoon depicting the meaning of each sentence formed. Only direct sentences with a simple grammar and unambiguous action can be understood by the program, although several sentences can be combined into a longer story. Overall, it provides a very amusing format for introducing simple stories.

Story Maker
Andee Rubin
Bolt Beranek, and Newman
10 Moulton Street
Cambridge, MA 02138
(617) 491-1850

Story Maker is a reading and writing activity appropriate for elementary, junior high, and high school age youngsters. It is comprised of a set of three activities: *Story Maker*—children construct complete stories by making choices among pre-written story parts; *Story Maker with Goals*—a goal is set for the youngster by the computer, and an evaluation of the resulting story takes place; *Story Maker Maker*—the youngster adds his/her own part to the story. The newly created part is then permanently stored with the other story parts so that the next child who uses *Story Maker* can use it in a story. The cost is $30.00.

Suspect Sentences
Ginn and Co.
191 Spring St.
Lexington, MA 02173

This program employs a game format for two or more participants, who need not be present simultaneously. One person inserts his or her forged sentence into a paragraph taken from a professionally written work. Another person then tries to guess which sentence is the forgery. The computer scores the accuracy of each guess and prompts the participants to discuss the insertion and its detection. Space is available for up to forty different passages. This program provides a very imaginative example of how computers can be used to develop complex skills such as awareness of literary style.

Games

Games can be highly motivating as well as challenging and interesting. Here are some examples. (See nearly any popular computing or educational computing magazine for descriptions and reviews of many others.)

Gertrude's Puzzles
Gertrude's Secrets
The Learning Company
4370 Alpine Road
Portola Valley, CA 94025

Snooper Troops
In Search of the Most Amazing Thing
Spinnaker Software
215 First Street
Cambridge, MA 02142

Logo

Apple Logo
Apple Computer, Inc.
20525 Mariani Avenue
Cupertino, CA 95014

Atari Logo
Atari Inc.
1265 Borregas Avenue
Sunnyvale, CA 94086

Commodore Logo
Commodore Business Machines, Inc.
1200 Wilson Drive
West Chester, PA 19380

DR Logo
Digital Research Inc.
P.O. Box 579
160 Central Avenue
Pacific Grove, CA 93950

Krell Logo (Apple)
Krell Software Inc.
1320 Stony Brook Road
Stony Brook, NY 11790

SmartLOGO
Coleco
200 Fifth Avenue
Suite 1234
New York, NY 10010

Terrapin Logo (Apple)
Terrapin Inc.
380 Green Street
Cambridge, MA 02139

TI Logo II
Texas Instruments Inc.
P.O. Box 53
Lubbock, TX 79408

Mathematics Software

The following mathematics software packages are cited in Chapter 4. (See *Computers in Teaching Mathematics*, Addison-Wesley Publishing Company, for descriptions and sources of many more.)

Green Globs
Conduit
P.O. Box 388
Iowa City, IA 52244

Darts
Control Data Publishing
P.O. Box 261127
San Diego, CA 92126

Arcademic Skill Builders
Developmental Learning Materials
One DLM Park
Allen, TX 75002

Speed Up Your Algebra
Eugene A. Herman
Department of Mathematics
Grinnell College
Grinnell, IA 50112

Plot
The Micro Center
Department S-42
P.O. Box 6
Pleasantville, NY 10570

Golf, The Jac Game
Milliken Publishing Co.
1100 Research Boulevard
St. Louis, MO 63132

SemCalc
Sunburst Communications
Room M5
39 Washington Avenue
Pleasantville, NY 10570

VisiCalc
Visicorp
2895 Zanker Road
San Jose, CA 95134

Software Directories

For a listing of the numerous software directories available see *Practical Guide to Computers in Education*, Addison-Wesley Publishing Company. Those listed below contain specific references to software for special needs.

Programs for the Handicapped
Prentke Romich Co.
8769 Township Road 513
Shreve, OH 44676
(216) 767-2906

This catalog lists software developed for the handicapped.

The SpecialWare Directory
LINC Resources, Inc.
1875 Morse Road, Suite 215
Columbus, OH 43229
(614) 263-2123

This directory lists instructional, administrative, professional, and evaluation/testing software for special education. A 28-page index provides curriculum and subarea access to both regular and special education software companies. A single copy is $13.95.

Swift's Educational Software Directory
Sterling Swift Publishing Co.
7901 South I-35
Austin, TX 78744
(512) 282-6840

This directory contains a section devoted to software for special needs.

HARDWARE

Perhaps the most dramatic impact of computers on special needs populations, has been the myriad hardware developments that are beginning to "normalize" the lives of people whose handicaps have heretofore severely restricted them. In this book we have discussed some of the more promising of these hardware developments as they impact the education of children with special needs. In this section we list sources of these and other hardware appropriate for use with handicapped individuals. Again, the listing is incomplete. New products come on the market daily, prices decline dramatically, and research and development continues at an impressive pace.

Despite the plethora of hardware developments, however, it is not at all clear that the lives of most handicapped individuals have been significantly altered. The high costs of much hardware and the low availability of trained persons to make the appropriate adaptations for a particular handicapped individual remain roadblocks for most. Nevertheless, the rich possibilities suggested by the great variety of hardware listed below is encouraging.

For convenience, the devices below have been divided into those which offer primarily alternative 'outputs' and those which provide primarily alternative means of "input."

Output Devices

For an in-depth review of the following speech output devices see Volume 2, No. 7 of the magazine, *Personal Computer Age*.

Echo GP Voice Synthesizer
Computer Gold
Micro Systems
2417 Coleshire Drive
Plano, TX 75074
(214) 596-2598

Echo PC Speech Synthesizer
Street Electronics Corp.
1140 Mark Avenue
Carpinteria, CA 93013
(805) 684-4593

Supertalker II
Mountain Computer
300 El Pueblo Road
Scotts Valley, CA 95066
(408) 438-6650

Type-'N-Talk
Votrax Division
Federal Screw Works
500 Stephenson Highway
Troy, MI 48084
(313) 588-0341

A Computer Terminal for Blind People
UFE Incorporated
1850 South Greenley Street
P.O. Box 7
Stillwater, MN 55082
(612) 439-1561

The American Federation for the Blind, in a project funded by the National Science Foundation, has undertaken the task of developing a full-page, paperless, transitory, tactile graphic Braille display. A mechanical matrix is used to raise or lower pins thus enabling a symmetrical pattern to be created that could then be used to generate Braille text and/or graphics.

Delphi and **Talktex**
Bill Duffy
RFD 1
Mashpee, MA 02649
(613) 477-3834; (617) 491-3393

Talktex is a combination voice terminal and voice output computer system.

ECHO II Speech Synthesizer
Laureate Learning Systems, Inc.
1 Mill Street
Burlington, VT 05401

This plug-in unit converts most correctly spelled words to speech output. The user may select pitch level and rate, and have words said aloud or spelled letter by letter. Punctuation and numbers can also be pronounced. The cost is $149.95.

Kurzweil Reading Machine for the Blind
Bernice A. Broyde, Marketing Director
185 Albany Street
Cambridge, MA 02139
(617) 864-4700

The KRM reads printed materials—books, magazines, letters, and reports—out loud. The electronic voice allows the blind student or worker access to material that would otherwise be unavailable. The KRM can also be attached to video terminals to provide speech output.

Large Print Computer (LPC)
Visualtek
1610 26th Street
Santa Monica, CA 90404
(213) 829-3969

Designed expressly for the partially sighted, the enlarged display letters are an automatic feature with no special program required. The LPC can be used with a variety of peripherals, such as a printer or joystick.

Micro-Communicator
Grover & Associates
7 Mt. Lassen Drive
San Raphael, CA 94903

Without the use of expensive peripherals, the Micro-Communicator transforms the Apple II/II+ into a communications device for handicapped individuals. A computer, single disk drive, and TV are the required hardware. With the use of a mouthstick, a single keystroke will enable an entire sentence to be displayed on the TV screen. A built-in vocabulary exceeding 1600 words allows longer messages to be constructed. There are two versions, child and adult. The cost is $45.00.

The Parrot
Research in Speech Technology, Inc.
P.O. Box 499
Ft. Hamilton Station
Brooklyn, NY 11209
(212) 259-4934

The Parrot, a plug-in speech module, is for use with the Timex/Sinclair Computers. It is capable of generating all sixty-four phonemes of spoken English. The user may combine these sounds in any order and can generate words, phrases, sentences, and sound effects. The cost is $89.95.

Phonic Mirror Handivoice
HC Electronics
250 Camino Alto
Mill Valley, CA 94941
(415) 383-4000

This handheld electronic voice synthesizer can produce virtually any word in the English language. The speech output is based on prestored sounds, words, and phrases which can then be combined into sentences. Sentences may be recalled and repeated. Handivoice operates on rechargeable batteries and is lightweight and portable.

Superphone
Prentke Romich Company
8769 Township Road, 513
Shreve, OH 44676
(216) 567-2906

The Superphone provides nonspeech communication and computer access for the physically handicapped or the hearing impaired. Transmissions are made via the same frequencies used by the deaf TTY network. The user may originate a call from the Superphone to any ASCII system equipped with an answer modem. The cost is $439.00.

TEL-AIDE
George A. Olive
Applied Micro Systems, Inc.
P.O. Box 832
Roswell, GA 3007
(404) 475-0832; (404) 371-0832

TEL-AIDE is a communications system that enables the deaf to communicate with anyone having a Touchtone phone, and eliminates teletype devices that limit the number of people with whom a deaf person can speak. TEL-AIDE runs on the Apple II Plus, which can be hooked into the phone using a modular plug. Using the Telephone Tone Code (TELTOC), individuals who wish to talk press the touchtone phone and the message is printed out on the computer printer.

Vidvox
Yvonne Russell
Sensory Aids Foundation
399 Sherman Avenue, Suite 12
Palo Alto, CA 94306
(415) 329-0430

Vidvox is a speech recognition device that will convert the spoken word to a printed representation (not traditionally spelled words). It may augment a hearing-impaired individual's access to speech.

Viewscan Text System (VTS)
Sensory Aids Corporation
Suite 110
White Pines Office Centre
205 W. Grand Avenue
Bensenville, IL 60106
(312) 766-3935

The VTS enables the visually impaired to "read" printed text with a miniature hand-scanned camera, store text on micro cassette tapes, and print out text via a miniature printer. The system is programmable in BASIC and has the ability to receive information from computer databases with the addition of a phone modem. The VTS is battery operated, no larger than a standard briefcase, and is completely portable. The cost is $6,320.00.

TURTLES

A turtle performs as a robot. It allows the user to program it to perform functions specified by the user (turn, move, draw). For the special needs user it encourages exploration of the environment and can provide a means of expression for those with communication handicaps. Two sources for turtles are:

Tasman Turtle
Harvard Associates
260 Beacon Street
Somerville, MA 02143
(617) 492-0660

Topo
FRED
Androbot, Inc.
101 East Daggett Drive
San Jose, CA 95134
(408) 262-8676

Input Devices

Keyguard for the Apple Computer
Prentke Romich Co.
8769 Township Road 513
Shreve, OH 44676
(216) 567-2906

This keyguard fits over the Apple keyboard. This adaptation is useful for some individuals whose erratic movements interfere with their typing accuracy. The user may use one finger, mouth stick, or headpointer.

Metasoft Speech Recognition Card
Christine Dhionis
3 Peter Circle
Marblehead, MA 01945
(617) 631-8968

The speech recognition card allows a student to use the computer without relying on the keyboard. It recognizes a small vocabulary of spoken words and phrases and responds to a variety of voices and pronunciations. The system adapts easily to the individual without special set-up procedures.

Shadow/VET
Prentke Romich Co.
8769 Township Road, 513
Shreve, OH 44676
(216) 567-2906

Shadow/VET is a voice recognition system that allows the user to control the Apple by speaking into a microphone. It can be used to access standard word processing software and is compatible with a variety of computer languages.

Voice Input Module for the Apple II
MCE Inc.
157 South Kalamazoo Mall
Kalamazoo, MI 49007

The VIM permits individuals to speak isolated words or phrases to their computer and thereby control any Apple II program. For an additional cost, the Apple IIe version is available.

Zygo Scanwriter
Zygo Industries, Inc.
P.O. Box 1008
Portland, OR 97207
(503) 297-1724

With this keyboard replacement, the user is presented with a matrix of choices, each marked by an LED. However, the user must be able to control a switch of some type. A scanner lights each LED cell-by-cell until the user responds by closing the switch at the appropriate time. Once a selection has been made, the code for that cell is transferred to the microcomputer as though it had been typed.

GAME PADDLES

Apple Computer, Inc.
20525 Mariani Avenue
Cupertino, CA 95014
(800) 538-9696

Coleco
200 Fifth Avenue
Suite 1234
New York, NY 10010
(212) 242-6605

Compu Sense
P.O. Box 18765
Wichita, KS 67218
(316) 263-1095

The Keyboard Company
7151 Patterson Drive
Garden Grove, CA 92641
(714) 891-5831

Tech Designs, Inc.
3638 Grosvenor Drive
Ellicott City, MD 21043

T G Products
1104 Summit Avenue
Suite 110
Plano, TX 75074
(214) 424-8568

GRAPHICS TABLETS

Apple Computer, Inc.
20525 Mariani Avenue
Cupertino, CA 95014
(800) 538-9696

Chalkboard, Inc.
3772 Pleasantdale Road
Suite 140
Atlanta, GA 30340
(800) 241-3989

Hewlett-Packard
1820 Embarcadero Road
Palo Alto, CA 94303

Koala Technologies Corp.
4962 El Camino Real
Los Altos, CA 94022
(415) 964-2992

JOYSTICKS

Apple Computer, Inc.
20525 Mariani Avenue
Cupertino, CA 95014
(800) 538-9696

Coin Controls, Inc.
2609 Greenleaf Avenue
Elk Grove, Il 60007

Coleco
200 Fifth Avenue
Suite 1234
New York, NY 10010
(212) 242-6605

Hayes Products
1558 Osage Street
San Marco, CA 92069
(619) 744-8546

The Keyboard Company
7151 Patterson Drive
Garden Grove, CA 92641
(714) 891-5831

Micro Stand, Inc.
2000 South Holladay
Seaside, OR 97138
(800) 547-2107

Mimco Stick
1547 Cunard Road
Columbus, OH 43227

LIGHT PENS

Data-Rite Industries
P.O. Box 976
Spokane, WA 99210
(800) 541-9001

Design Technology
7760-B Vickers Street
San Diego, CA 92111
(619) 268-8194

3-G Company, Inc.
Rte. 3
Box 28A
Gaston, OR 97119

Tech-Sketch, Inc.
26 Just Road
Fairfield, NJ 07006
(800) 526-2514

T G Products
1104 Summit Avenue
Suite 110
Plano, TX 75074
(214) 424-8568

TOUCH SENSITIVE SCREENS

Interaction Systems, Inc.
24 Munroe Street
Newtonville, MA 02160
(617) 964-5300

Touch Technology, Inc.
3 Church Circle
Annapolis, MD 21401
(301) 269-8838

Index

Access, 24–28, 54, 88
Achievement test, 152, 161
Adventure games, 4, 61, 64–70, 75, 80–81
ALS (Lou Gehrig's disease), 50
Anecdotal records, 165
Arcademic Skill Builders, 33, 94
Arithmetic, 54, 89, 93, 95, 156 (*see also* Mathematics)
    addition, 89, 97, 114, 117
    counting, 160
    decimals, 89
    division, 89–90, 93, 97
    fractions, 89, 93, 97, 100, 158
    multiplication, 89, 93, 97
    numerals, 93, 164
    subtraction, 89, 97
Art, 57, 124–125, 128–132, 146
Arthritis, 50
Arthrogryposis, 50
Artificial intelligence, 34, 40
Assessment, 146, 149–167
Athetosis, 142, 179
Attribute blocks, 115
Authoring systems, 94
Autistic, 55, 65, 134–135, 137, 144, 186
Autonomy, 123–147

BASIC, 170–172
Behavior modification, 145
Bilingual, 158, 161
Binet, Alfred, 152
Biography, 75
Blinking lights, 180

Blindness (*see* Visual impairment)
Bliss symbols, 57
Braille, 29
Braille keyboards, 106
Branched-plot books, 75
Button box, 135, 160, 164

CAI (*see* Computer assisted instruction)
Calculator, 90
California School for the Deaf, 94
CARIS (Computer Animated Reading Instruction System), 51–52, 71
Cerebral palsy, 1, 13–17, 24, 28, 50, 55, 66, 83, 139, 160
Children's thinking, 156–161
Chronic ear infections, 55
Color blindness, 6
Color TV sets, 180
COMCAL, 189
Commack, Long Island school system, 189–191
Communication, 28–29, 45–85, 88, 137, 161
    needs, 96
    skills, 68
Computer aided speech recognition, 53
Computer assisted instruction (CAI), 7, 94, 95, 113, 170, 171
Computer-based language activities, 58
Computer-based testing, 167
The Computer Chronicles Newswire, 82
Computer literacy, 90, 171
Computer mail, 11–12, 53, 57, 59, 82, 128

Computer networks, 40, 53
Computer science courses, 189
Computer tools, 81–85
Computers and access, 34–40
Computers and motivation, 32–34
Computers and physical handicaps, 35–40
Computers and the deaf, 40–41
Computers in the workplace, 89
Cooperation, 65
CP (*see* Cerebral palsy)
Creativity, 146
Curriculum strategies, 159
Curriculum tools, 167
Cursor, 180

Darts, 97–98, 158
Deaf (*see* Hearing impairment)
Disks, 154, 176
Disk drive, 176, 181
Drawing tool, 109
Drill and practice, 58, 61, 71, 73, 79–80, 93–95, 113, 160, 170
Dungeons and Dragons, 61, 64
Dyslexic, 26, 145, 179

Education of All Handicapped Children Act (P.L. 94-142), 164
Educational plan, 57, 60, 128
Educational Testing Service, 94
Electronic mail (*see* Computer mail)
Electronic scratch pad, 106–107
Electronic workbooks, 71
Emanuel, Ricky, 134, 144
Emotional/behavioral difficulties, 23, 55–56, 114, 184
  aggressive behavior, 126
  antisocial behavior, 126
  assertive behavior, 127
  attention deficit, 96, 102, 103, 160, 184
  low frustration tolerance, 103
  resistance, 126
English
  as a second language, 82
  Standard, 65
  structure, 67
  syntax, 67
Epileptic, 180
Error analysis, 156–157

Exploration, 63, 79, 113
  in language learning, 72
  in science, 125
Eyeglasses, 193
Eyestrain, 177

Feedback, 163
Florida School for the Deaf, 94
Food and Drug Administration Bureau of Radiologic Health, 180

Gallaudet College, 52
Game paddles, 182
Game-paddle input port, 184
Games, 57–58, 79, 95–106, 113, 180
Gertrude's Puzzles, 115
Goldenberg, Paul, 78, 135
Graduate Record Examination, 149
Graphics, 123, 129–130, 146
Graphics printer, 129
Graphics tablet, 1–2, 16, 34, 36, 142, 183–184
Green Globs, 104

Hand coordination, 183
Handwriting, 53
Hawkins, David, 108
Head wand, 106, 139, 181, 183
Hearing aids, 193
Hearing impairment, 12–13, 29, 51–53, 55, 82–83, 88, 94, 113, 117–118, 128, 187, 190
High spinal cord injury, 50, 186
Hunt the Wumpus, 158–159
Hutchens, Marcia, 165
Hyperactive, 102, 114, 144

Indian Hollow Elementary School, 190
Individual educational plans (IEPs), 20, 164–167
Individualized instruction, 93, 97, 154
Individually administered tests, 150
Informal assessment, 152
Interactive Text Interpreter (ITI), 76, 80

Jar Game, The, 98
Joy stick, 34, 38, 142, 164, 182–183, 190

Keyboard, 176, 181
Keyboard guard, 187
Kraus, William, 98
Kruteskii, 93

Language arts, 45–85, 190
Language-as-currency model, 60–61, 63–71, 81
Language-as-subject matter, 60–61, 63, 71–81
Language handicaps, 94, 138
Language-learning environments, 59–63
Learned passivity, 125–128, 140–141
Learning disabilities, 23, 87, 94, 96–97, 113, 126, 163
Learning style, 150
Lemonade, 117
Levin, Jim, 82
Light pens, 34–35, 106, 164, 182, 184
Light-sensitive probe, 184
Logic, 90, 113, 190
Logo, 59, 62, 74, 80, 109–114, 129–130, 132, 159–160, 163, 172, 193
Logo Laboratory at M.I.T., 135, 159
Los Angeles Unified School District, 94

Mainstreamed, 96, 114, 128, 190
Make-a-Monster, 102
Management, 167
Massachusetts Institute of Technology (M.I.T.), 135, 159
Matching, 109, 160
Mathematics, 87–120, 137, 145–146, 190
  algebra, 90, 104, 113
  applied, 89–90
  as critical filter for women and minorities, 88
  as vocational prerequisite, 87–88
  calculus, 132
  charts, tables, and graphs, 90
  circles, 103, 109, 112
  computation, 91
  coordinate axes, 104
  curriculum, 118
  estimation, 90–91, 93, 109, 112, 114–115, 160
  geometry, 90, 93, 103, 108, 113, 129
  graphing, 93
  hierarchy, 90
  laboratory, 114
  literacy, 89
  measurement, 90, 112, 115
  new math, 192
  number line, 97
  number theory, 113
  play, 95, 108–113
  plot, 108
  polygons, 109, 112
  prediction, 90
  probability, 93, 98, 100, 113
  problems, 161
  tools, 95, 106–108
  topology, 113
  trigonometry, 109
  worlds, 109
Medical insurance, 193
Meichenbaum, Donald, 144
Mentally retarded, 55, 77, 94
Mercury switch, 184
Meteor Multiplication, 33
Microworld, 95, 109, 115
Mill Neck Manor School for the Deaf, 190
Modem, 41
Motivation, 22–24, 123–147
Motor control, 163
Muscular distrophy, 50
Music, 57, 124–125, 128–132, 145–146
Music synthesizer, 16
Mutism, 56
My New Life, 126

National Assessment of Educational Progress, 89, 119
National Council for Radiation Protection and Measurements, 180
National Council of Supervisors of Mathematics, 90
Neurological damage, 77
New England Journal of Medicine, 180
New York State Regents Examination, 152
Noise-operated switch, 186
Normal curve, 150

Office of Education, 93
Optical character recognition (OCR), 84
Optical sensors, 184

Orff instruments, 128
Overlay, 188

Papert, Seymour, 109
Paintings Library, 100
Paralyzed, 186
Perceptual-motor problems, 184
Peripheral devices, 35
Personal dictionary, 155
Phonemes, 77
Physically handicapped, 35–40, 87–88, 94, 96–97, 107, 113, 118, 141, 164, 190
Physics experiments, 143
PLATO, 100, 102
Play, 109, 113, 160
Portable computers, 178–179
Portable speech recognition devices, 40
Practical Guide to Computers in Education, 169, 175
Prelingual deafness, 25
Printer, 154, 177
Problem-solving, 89–90, 94, 104, 113, 146, 158, 160
Processor, 176
Programming, 109, 139, 170–173, 189, 191
Project CAL (Computer Accelerated Learning), 189
Prostheses, 34–35, 106, 139, 163, 167
Proximity-sensing technology, 185
Psychotic, 135
Punctuation, 71, 76, 153

Quadriplegic, 106

Radiation, 179
Reading comprehension, 75
Reading disability, 56
Reading machines, 54
Reasoning, 158
Record-keeping, 165, 167
Recreation, 57
Reinforcement, 94
Remediation, 151, 153
Report card, 165
Riel, Margaret, 25, 27, 126
Reye's syndrome, 186

Robot turtle (*see* Turtle)
Role of the teacher, 118

Safety, 176
Seizures, 176, 180
Self-discipline, 127
Self-evaluation, 150
Self-governance, 127
Self-image, 126
Self-regulation, 144
SemCalc, 107
Sentence construction, 72–74
Shanahan, Dr. Dolores, 189
Sharples, Mike, 74
Sign language, 12, 57
Simulations, 63, 65, 80–81, 113, 115, 117, 158, 167, 190
Snooper Troops, 65
Social interaction, 132
SOPHIE, 48–49, 53, 60, 63–65, 72, 80
Sound-input device, 186
Spastic quadriplegia, 193
Spasticity, 179
Spatial information, 93, 182–183
Special Technology for Special Children, 82
Speech, 146, 163
    defects, 56, 77, 97, 106
    digitizers, 34 (*see also* Voice synthesizer)
Speechread, 12, 53, 83
Speechreading aid, 83
Speech-to-text conversion (*see* Voice recognition programs)
Speed Up Your Algebra, 107
Spelling, 153–154, 179
    checker, 145
    difficulty, 82
    drills, 71
    lessons, 58
Spina bifida, 27
Staff development, 189
Standardized tests, 90, 149–150
Stanford Achievement, 163
Statistics, 108, 113
Storymaker, 62, 74–76, 80–81
Stroke, 50, 59
Stuttering, 56

Suppes, Patrick, 93
Suspect Sentences, 76, 80–81

Talking and writing aid, 181
Talking calculators, 54
Talking computer terminal, 54, 60
Teacher Planning System, 167
Teaching machines, 192
Testing, 149–151, 162
Textman, 62
Text-to-speech device, 15, 84
Text-editing, 153, 165
Theremin, 185
Time-shared computer system, 189
Title IV-C, 189
Touch-sensitive screens, 34, 37, 106, 182
Touch-sensitive switch, 186
Turtle, 109, 114, 129–130, 132, 134–135, 144, 146, 160, 164
TV, 177, 180
Typing, 153

University of Edinburgh, Scotland, 74

van Lint, June, 126

VDT, 35
Video ping-pong game, 58, 60, 142
VisiCalc, 108
Visual impairment, 26, 53–55, 87–88, 96–97, 145
Vocabulary builders, 71
Vocabulary training, 172
Voice-activated computers, 34
Voice input, 164
Voice recognition programs, 34, 53, 186
Voice synthesizer, 4, 34, 183

Weir, Sylvia, 134, 144, 159
Wheelchair, 179, 186, 190, 193
Wheelchair-mounted computer, 15, 178
Wilson, Kirk, 167
Word processing, 60, 63, 76, 81, 106, 145, 171, 179, 190
Word search puzzles, 58, 61
Work environment, 177
Writer's Assistant, 76
Writing, 146
  laboratory, 81
  skills, 153
  stories, 74–77, 125